BUSINESS ISSUES, COMPETITION AND ENTREPRENEURSHIP

NOISE AND ERGONOMICS IN THE WORKPLACE

Business Issues, Competition and Entrepreneurship

Additional books in this series can be found on Nova's website under the Series tab.

Additional e-books in this series can be found on Nova's website under the e-book tab.

BUSINESS ISSUES, COMPETITION AND ENTREPRENEURSHIP

NOISE AND ERGONOMICS IN THE WORKPLACE

PAULO HENRIQUE TROMBETTA ZANNIN
EDITOR

nova publishers
New York

For permission to use material from this book please contact us:
Telephone 631-231-7269; Fax 631-231-8175
Web Site: http://www.novapublishers.com

Library of Congress Cataloging-in-Publication Data

ISBN: 978-1-62417-573-2

Published by Nova Science Publishers, Inc. † New York

*This book is dedicated to the memory of my parents Altayr and Belony,
to my wife Carolina, my twin children Beatriz and Vitor,
to Großmutter Irene Schmidt-Endrulat, and to my academic and scientific
mentors, Prof. Dr. rer. nat. Manfred Heckl (in memoriam)
and Prof. PhD Samir Nagi Yousri Gerges*

CONTENTS

PREFACE

This book addresses two extremely important aspects of workplaces, namely, *noise and ergonomics*. Its thirteen chapters present and discuss *theoretical and practical* issues involving *noise and/or ergonomics*, covering a wide and diverse spectrum of working environments that directly affect thousands of workers around the world. The design of the environment is discussed through the relationship between ergonomics and architecture, revealing that the efficient integration between these factors allows us to outline boundaries in the search for solutions to the various specificities of the organization and its different workplaces. The analytical techniques used throughout the book are widely diverse and include *in situ* acoustic measurements, computer simulations and analysis of noise perception.

The various working environments evaluated herein include *schools*, *universities*, *health care facilities*, *fitness centers*, *factories*, *open plan offices*, and *urban buses*. The work environments evaluated in this book can be considered universal, since they are found in hundreds of countries. Therefore, the topics presented and discussed here, maintaining the proper distance and relativity specific to each country, can help to solve and/or understand local problems.

This book was written by teachers, researchers and professionals working in the fields of acoustics and ergonomics, but with different backgrounds and areas of expertise. Hence, its approach is clearly multidisciplinary. The authors of the various chapters of this book are engineers, psychologists, ergonomists, registered nurses, architects, physicists, audiologists and physical education and sports specialists.

Several case studies are reported in this book. These are examples of real cases, in which field measurements were taken to evaluate the quality of the workplace from the standpoint of ergonomics or noise control. In some cases, interviews were also conducted to quantify the interviewees' perceptions of the ergonomic and acoustic quality of the workplace environment. Computer simulation techniques were also used in the quest for a healthier workplace environment.

This book will be of interest for use in theory and practice. We believe that numerous professionals, even those not specialized in noise or ergonomics, will find it useful. The target audience for this book includes mechanical engineers, civil engineers, ergonomists, designers, architects, audiologists, health care professionals, school builders and administrators, undergraduate and graduate students. This book presents new and significant research in the field of noise and/or ergonomics in the workplace.

Last but not least, the Editor would like to thank all the authors for dedicating their time, efforts and expertise to writing their respective chapters, and especially Ms. Beatrice Allain for her excellent work in translating all the chapters from Portuguese into English, with the exception of Chapter 1, which was originally written in English.

Curitiba
September 21, 2012

In: Noise and Ergonomics in the Workplace
Editor: Paulo Henrique Trombetta Zannin

ISBN: 978-1-62417-573-2

Chapter 1

DESIGNING THE POSITIVE PUBLIC SCHOOL ENVIRONMENT: A BRAZILIAN PERSPECTIVE

Doris Catharine Cornelie Knatz Kowaltowski**, *Marcella Savioli Deliberador and Paula Roberta Pizarro Pereira

Department of Architecture and Construction, School of Civil Engineering, Architecture and Urban Design, University of Campinas (UNICAMP), Campinas, São Paulo, Brazil

ABSTRACT

Public education has always been an important instrument to promote social progress. Schools need to address issues that impact society and make sure that future citizens are capable of meeting the challenges that lie ahead. The quality-learning environment is composed of intelligent students, excellent teachers, appropriate pedagogy and an engaging, comfortable and attractive school building. To attain a quality built environment the design process needs structure, rigor and rich and varied data on architectural design elements and their impact on environmental comfort, functions and human behavior. The extensive literature on school architecture presents a variety of aspects that influence the learning process, not least functional elements. Thus, aspects of space per student and school size, that affect densities and feelings of crowding, are important; as are classroom shapes and layouts, color schemes and wayfinding issues in school buildings. The quality and distribution of equipment and furniture to support the variety of learning activities of recommended contemporary pedagogies also need to be discussed in school building design processes. In many countries the traditional school building and classroom layout for up to 45 students are still the norm and this is not yet questioned in relation to the rich insights coming from research on the learning environment. This chapter presents data from a continuing study on the public school environment as found in the State of São Paulo, Brazil. The local school building design process was characterized and shown to lack a detailed briefing phase to discuss the issues shown above. Some opportunities for change were identified and the incremental introduction of improvements to local schools should be explored.

* E-mail: doris@fec.unicamp.br.

1. Introduction

Public education has always been an important instrument to promote social progress and in developing countries, like Brazil, the importance of education cannot be underestimated. Schools need to address issues that impact society and make sure that future citizens are capable of meeting the challenges that lie ahead. Schools should set an example to adequately represent the desired values of a specific time and place. These values can be externalized through curriculum contents, teaching methods and an engaging, comfortable and attractive built environment, expressed through school architecture that is appropriate (Samad and Macmillan, 2005). These elements represent what the educational system stands for in a specific locality and period.

The primary issues to be considered in a school design process are, no doubt, related to functional aspects of the building to embrace the educational activities indicated by the pedagogy and educational system adopted. These issues change over time and are place specific. For instance, one of the more urgent issues of social progress today, on a global scale, relates to environmental questions and the means of implementing a society with more sustainable attitudes (Ford, 2007). Schools have a key role to play, through their educational principles, expressed as well through a sustainable architecture.

The construction industry is known to be a major consumer of natural resources and buildings-in-use impact the environment, producing various pollutants and large quantities of energy and water are spent to satisfy user needs. Educational institutions are part of this reality and high performance schools are seen as responsible alternatives (CHPS, 2009). Sustainable school buildings, or green schools, are seen as providing healthy environments, which support educational trends and aid in increasing academic performance of students (Chase, 1995). Architecture of this kind should cost less to operate, and contribute in saving water and energy of a community. In Brazil sustainable proposals are best based on the principles of bioclimatic architecture, since most of the territory of the country has favorable tropical or sub-tropical climatic conditions to support such designs.

Not withstanding these new demands exerted on schools, an educational environment in general should offer spaces that adequately accommodate academic and social activities as part of the pedagogy adopted and affect positively the social dynamics of a school community. These demands on the school environment are manifold, and a prime function of its architecture is to respond to these. A growing number of studies show that the quality of the physical environment influences users, their level of stress, physical and mental health, as well as a sense of self-esteem (Lackney, 1998, 2000; Maxwell and Chmielewski, 2008). Children spend a large amount of time during the day in schools and buildings with poor ventilation, inadequate lighting and acoustical conditions, as well as unfavorable heating or cooling systems have direct consequences on well-being factors and impact the larger environment as well (Lackney, 2006). Research concludes that well maintained, clean and safe buildings demonstrate higher levels of academic performance of pupils, then those from schools with broken windows, leaky roofs and dark and dingy classrooms (Upitis, 2004). For school architecture to influence user well-being positively and embrace its important role as an educational agent, the design process must be critically reviewed and adjusted so that professionals can respond to current demands within local constraints.

In Brazil the quality of public education has been under debate, especially in light of the unsatisfactory performance levels obtained by students in general tests. Most discussions center on the system of public education, curriculum and pedagogy, but should also touch on the adequacy of the physical environment to support necessary educational changes to increase performance levels (Gomes da Silva et al., 2009). The need for change can be corroborated by results of Post-Occupancy Evaluations (POE) of school buildings in Brazil, which show that problems related to environmental comfort are frequent and schools lack a variety of spaces to support a rich array of recommended educational activities, indicating that current design parameters require revision (Ornstein, 2005; Kowaltowski et al., 2001). A discussion on the design process of public schools in Brazil is presented here, with emphasis on the State of São Paulo, where school buildings are managed by a government agency called FDE (*Fundação para o Desenvolvimento da Educação*) (Deliberador, 2010). Opportunities to improve this process and therefore the product are identified and functional aspects of the school environment are detailed as design requirements to support an enriched design process.

2. A Recommended School Design Process

The literature on school buildings is vast, discussing tendencies in education and the architectural response to new teaching methods and their necessary technological support (Dudek, 2007; Nair and Fielding, 2005; Ford and Hutton, 2007). Architectural concepts are established to ensure a quality educational setting. These need to be adjusted to specific local conditions, to be effective. Architectural space can be considered the third teacher of the school environment, to support the educational environment of excellent teaching staff and the application of responsible curriculum and creative teaching methods.

According to the many studies, relating student achievement to environmental quality conditions, classroom acoustics directly affects communication and exterior noise is a common problem in urban area schools (Duran-Narucki, 2008; Boman and Enmarker, 2004; Crandell et al., 2005). Crowding in classrooms can create discipline problems that can cause noise and affect intelligibility of verbal communication.

Crowding and the physical layout of classrooms can also curtail desirable educational activities and groupings for specific learning experiences. The total school population can affect the individual student's role in the educational environment, offering more or less opportunities and responsibilities (Barker and Gump, 1964; Andrews, et al., 2002).

Wall and ceiling colors are shown to affect student's outlook on education (Tanner, 2000). The lighting condition of schoolrooms is another important factor for an appropriate learning environment. This aspect affects the legibility of information. Daylight is highly influential and considered important (Heschong Mahone Group, 1999; Heschong, 2003). Thermal conditions can affect directly the health of building users and adverse situations can cause apathy and even stress (Myhrvold et al., 1996). Thus, the four main aspects of environmental comfort (thermal, acoustics, lighting and functional space relationships) should be at optimum conditions to provide the background for a good learning experience in schools (Schneider, 2002).

Post-Occupancy Evaluations (POE) of school buildings are important sources to identify environmental quality, building performance pathologies and human response to physical conditions. They should be part of a school design process to avoid the repetition of errors. In countries with good minimum standards of environmental comfort, such studies point out that not all problems are eliminated by regulations and an appropriate environment depends on many interconnecting factors, which must be addressed in the design process (Bordass et al., 2006; HEFCE, 2006; Lackney, 2001; Ornstein, 1997; Sanoff, 2001b; Watson, 2003; Rashid and Zimring, 2008; Zimring et al., 2005).

Besides basic questions of comfort, school buildings need to address wider issues of architectural quality to make the school community feel at home and stimulated. Sanoff (2001a) emphasizes the image of a school as a priority aspect and presents, in his design methodology, principles of school building design. Thus schools should have stimulating environments, places for group learning, linked outdoor and indoor places, enriched public spaces like corridors, safety, spatial variety, flexibility, good access to resources, active and passive places, personalized space and extend the school environment to the community as a learning place.

The quality of school building design depends on known design criteria and professional knowledge and practice, as well as feedback from building performance assessments and design evaluation tools. An example of such a tool is DQI the Design Quality Indicator, which assesses the quality of buildings according to: Functionality (the arrangement, quality and interrelationship of spaces and how the building is designed to be useful to all), Built Quality (the engineering performance of the building, which includes structural stability and the integration, safety and robustness of the systems, finishes and fittings) and Impact (the building's ability to create a sense of place and have a positive effect on the local community and environment) (Gann et al., 2003).

For school buildings, the Pattern Language by Alexander et al., (1977) has been adapted by Nair and Fielding (2009) with 25 parameters to support the development of a quality school environment. These include specific patterns relating to the primary learning environment "the classroom", detailed in further patterns called "Campfire", "Watering Hole" and "Cave Space". A welcoming entry is important with student display space and a home base and storage areas for students are recommended. Details for science and arts laboratories and life skill areas are presented, as are special areas for art, music and performances in general. Physical education and fitness spaces are discussed. Casual eating areas are recommended. Some concepts such as transparency and the need for interior and exterior vistas, as well as dispersed technology are introduced with emphasis on indoor-outdoor connections. Furniture is not forgotten and soft seating recommended. Flexible spaces are important and designing for multiple intelligences is a major issue in school design.

Aspects of environmental comfort are included with patterns for daylight, solar energy and natural ventilation. The relation between learning and lighting, as well as color is then discussed and sustainable elements are seen as part of the 3rd teacher concept or the building as a 3D textbook. The architectural language of the building is asked to have a local signature, so that a school is capable of demonstrating its connection to the community. Finally a pattern to bring all the previous requirements together is included in this specific school design language. All patterns present a discussion of the specific issue and a graphic representation of the solution scope. The 25 patterns express major functional needs of schools through "if.. then" discussions that inspire, illustrate and direct the design solution search. These design patterns should fit into an overall design process, which for learning environments is a complex assignment. Simple and elegant solutions can be proposed, but never through simplistic ideas, uni-dimensionally resolved.

The design process, which supports the development of a quality school environment, can be called a reference design process (Figure 1).

Unlike traditional ways of designing, this process structures procedures for the design of high performance schools.

The quality of education in a broad way should be the basis of such a process. The physical environment is considered an essential partner to achieve productive learning experiences. The reference process adds a multidisciplinary team to the process and values assessment at various stages along a non-linear structure. Knowledge is thus acquired and a feedback loop established, avoiding the repetition of errors. Appropriate tools are introduced to simulate comfort levels and other important functions of a building, to avoid later as-built assessment problems.

Users, or potential users (teachers, parents, students, school officials and staff) should be involved in the decision-making process. The professional designer must learn to act with ethics and responsibility in serving the needs of others. Thus, user expectations must be guided towards a proper understanding of a design's response to needs, to avoid disappointment and dissolution.

A recommended design process should have an analytical data collection phase to support the briefing activities that follow. POE and case study data are essential to enrich the programming discussions, as is access to regulations and recommendations for environmental comfort standards (Moreira and Kowaltowski, 2009).

Many methodologies exist to support the briefing phase. The "Problem Seeking" method, by Penã and Parshall (2001) presents a rich array of concepts and issues to apply to this phase and the ISO 9699:1994 international code (1994) regulates this activity as well. Value systems should be discussed and priorities, desires and goals established (Hershberger, 1999). Hershberger's list of architectural values can be used to structure programming discussions. Many guidelines have been written and Herschberger, in 1999, expanded the three original principles of architecture by Vitruvius, *firmitas, utilitas* and *venustas*, to 8 value areas, denominated HECTEAS or "Test each" as a basis for the architectural design process, especially its briefing phase. These values and their related issues are:

1. Human: functional, social, physical, physiological and psychological
2. Environmental: site, climate, context, resources and waste
3. Cultural: historical, institutional, political and legal
4. Technological: materials, systems and processes
5. Temporal: growth, change and permanence
6. Economic: finance, construction, operation, maintenance and energy
7. Aesthetic: form, space, color and meaning
8. Safety, structural, fire, chemical, personal and criminal

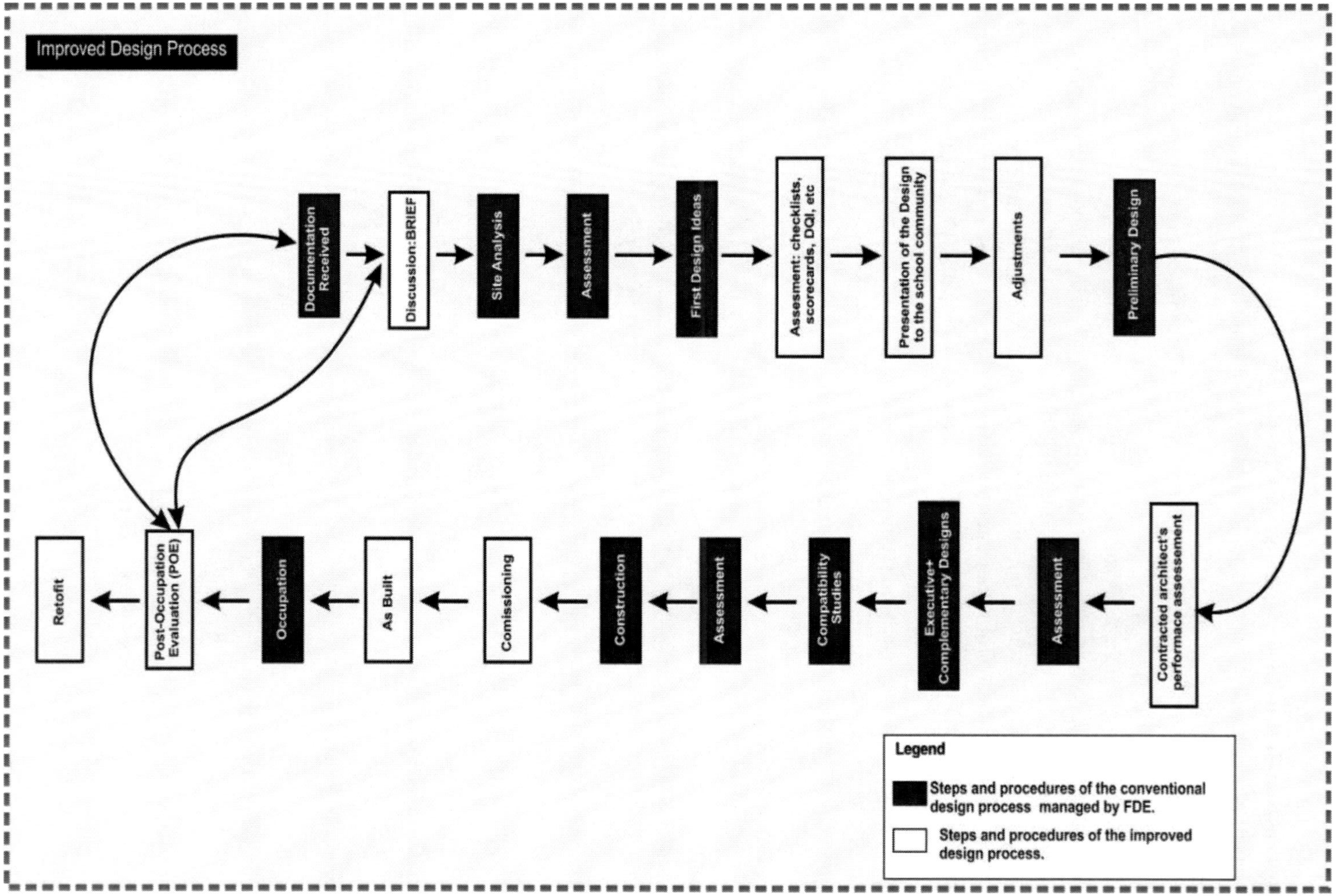

Figure 1. A reference design process structure.

Values stand as important reminders of issues that any building project should address and relate to Maslow's needs-pyramid (Maslow, 1943). Benedikt (2008) expands on these needs, stressing attention given to: survival (structural soundness, protection from climate, animals and projectiles); security (protection from trespass, seizure of persons or property by others, privacy and control over spaces, not just firmness is sought but assurance of this need); legitimacy (announcing social identity, establishing authority, laying claim to property, distinguishing peoples membership of different institutions and groups); approval (legal and value positive (aesthetic, social, economic additions to a neighborhood and making occupants look handsome, healthy and worthy); confidence (spontaneity, freshness in form, secure in its statements, élan, replacing less value with more) and freedom (of movement, opinion, space, flexibility, exclusion and privacy). Benedikt argues that value in architecture goes beyond the minimum or obvious incorporation of these needs. Values include the meeting of needs on the highest degree of satisfaction, through mindful and well-designed products or buildings. Through his theory, Benedikt (2008) provides an important model for value incorporation in building design and shows ways of achieving this through what he terms: persuasion using examples, encouragement and even if necessary flattery.

Once a value system is agreed upon, the design process (Figure 1) advances by analyzing the site and urban surrounding thoroughly, especially the physical characteristics and the local climatic and access conditions should be evaluated. Preliminary design solutions are then brought forward. At this stage an integrated design process should be introduced to assure that various specialists contribute at an early stage to design solutions (Figueiredo, 2009; Larsson, 2009; Aldrich, 2011). Such a design process consists of at least five stages: pre-design, design development, legal and contract documents, construction, facility performance and commissioning. The integrated process differs from the traditional way of procuring a new building or urban space development. The traditional process is linear and systems are considered in isolation, since consultants and team members with specialized knowledge are only involved when necessary. Integration on the other hand demands a whole systems approach and more team members are included throughout the process. Especially, at early stages of the design development decisions are made in teams with emphasis on the relationships of different systems as well as life-cycle costs and benefits (Aldrich, 2011). For school buildings, design options should be presented to the community and assessed in relation to economic, functional aspects, aesthetic and "buildability" perspectives (Wong et al., 2009). The preliminary design is then detailed and important simulations and evaluations performed to assess, in an integrated and multidisciplinary manner, comfort and energy efficiency conditions.

Assessing sustainability of new projects has become mandatory in many countries and for public schools in the State of São Paulo in Brazil the French system *Referentiel Technique de Certification - Bâtiments Tertiaires* (FCAV, 2007) was adapted in 2007 for local conditions called "Processo AQUA". For High Performance schools in the United States of America the Collaborative for High Performance Schools establishes a Best Practice Manual with checklists which include criteria such as (CHPS, 2009): site conditions (selection, transportation, storm water management, outdoor surfaces, outdoor lighting); water efficiency (water use budget, water for landscaping, water use reduction); energy efficiency (energy performance, natural ventilation, renewable energy, system testing and training, commissioning, energy management); materials use and choice (waste reduction and efficient material use, storage and collection of recyclables, site waste management, building reuse, resource reuse, sustainable materials recycled content, renewable materials, certified wood); indoor environmental quality (day lighting, indoor air quality, low-emitting materials, pollutant source control, construction indoor air quality management plan, minimum acoustic performance, improved acoustical performance, thermal comfort, ASHRAE code compliance, controllability of HVAC systems); district resolutions (policies, maintenance plans equipment performance, Green Power, transportation).

For local Brazilian conditions an evaluation method for comfort and functional issues was developed based on multi-criteria optimization (Graça et al., 2007). This optimization method can help designers in the decision making process at the preliminary design phase. Spatial configurations, which influence environmental comfort parameters for school buildings for the public school system of the State of São Paulo, Brazil, were analyzed and experts identified design solutions, which include recommended comfort variables. The application of the concept of optimization during the design process allows decision making to occur in a more rational manner, since design variables can be selected according to a qualification process. The method is restricted to four aspects of environmental comfort (thermal, acoustics, daylight and the functional relation of classrooms to service spaces). The use of a design method, which includes the concept of optimization, gives designers coherent arguments for decisions, making design proposal presentation and justification easier. Design proposal communication with users and clients is also improved. Optimization thus helps in the implementation of design solutions, avoids the adoption of inferior solutions and directs improvements of compromise solutions. The application of this method is seen thus as an important tool for the design of better school buildings and can also direct the introduction of improvements to existing schools. A preliminary design can also be checked for optimized use of available space. A tool called SPATE (2004) was developed in Brazil. This tool can be used for the evaluation of new designs or for space management in existing buildings. The basis of the tool is the timetable of educational activities according to teaching groups and their necessary resources and the available spaces according to typologies, and capacities. Such a tool is useful for school management, but it does not assess building design quality, only space quantities. Other design evaluation tools should also be applied. As mentioned previously the DQI for schools (Gann et al., 2003) should be used and the Comparative Floor plan Analysis (CFA) method by van der Voordt, Vrielink and Wegen (1997) can contribute by evaluating functional and specific educational criteria of a preliminary design. The CFA method is based on a comparison of architecture designs in relation to three key aspects: site, building, spaces. The analysis seeks to understand why differences in design solutions occur, by linking design alternatives to data from POE on the underlying arguments and experiences of users. This gives insight into relevant points of decisions and (dis)advantages of design variants for use and perception. This process occurs by interactive and iterative actions and should pass through the three levels of order (site, building and spaces). On the one hand, this analysis is guided by hypotheses, questions and notions of participant designers and their clients, data from POEs, review of literature and so on, and also the evaluator's own hunches. On the other hand, the floor plans themselves generate ideas and hypotheses, which can be checked through further sources. As a result spatial architectural choices become more understandable, recognizable and debatable, therefore clarity and transparency can be achieved. The CFA method can be applied in several ways:

- Within one functional building type, floor plans per se can be compared with each other with regard to a number of physical properties related to social characteristics such as confidentiality and privacy, communal use of space, zoning, etc..
- Comparisons between different functional building types such as schools, nurseries, libraries and health centers are also possible.
- Comparisons can be made synchronously (looking for cultural or regional differences, for example) and diachronically (looking for developmental changes).
- Results can be used to develop or to improve design criteria in general, such as guidelines, standards and users' requirements.

The CFA method is seen as an important addition to the traditional socio-scientific research methods, such as interviews and observations, and design research purely focusing on developing formal typologies (van der Voordt, Vrieling and Wegen, 1997). Using CFA with POE shows that behavioral aspects can be coupled directly to design variants, while sufficient freedom remains for making conscious choices on one's own design (the designer's own), or for the most suitable. Moreover, this integral approach gives designers an opportunity to support their way of looking at buildings, through floor plans methodologically. CFA is, of course, confined to those spatial means that can be read from floor plans. The influence of aspects like color, use of materials, the third dimension, furnishings and fittings on the organization and the actual behavior of people and vice versa cannot be measured in this way. CFA is therefore meant, above all, as preparation for and integration into POE studies.

The "Balanced Scorecard" method by Wong et al. (2009) can contribute as well in the assessment of designs. Four important design factors are set out as evaluation goals: aesthetics, functionality, "buildability" and economics. The method then details how such goals should be measured. Thus, an attractive design is measured through its volumetric proportions, patterns of design elements and colors, while functionality is based on convenience, accessibility of the various parts of a building, shape and size of spaces and their fit for purpose and the aesthetic consistency between building components and appearance. Construction efficiency is a specific goal and should be optimized with other design objectives, such as aesthetics and functional convenience. Economic concerns as well must be related to the various other design factors but also need to be assessed against profitability goals. In a school building design process, once the preliminary design has been thoroughly checked and assessed it should be approved by the client and participating users. Many adjustments are usually introduced during this first design phase. Once approved, specifications and final drawings are produced. The construction cycle can then begin. The multidisciplinary team should accompany this phase to make sure that execution is true to design intentions and small corrections can be smoothed out for some minor problems observed.

The commissioning phase is of vital importance to implement fine-tuning of the building, train staff and users in the proper operation of its infrastructure (Holtz, 2004). A minor retrofit phase may follow and a thorough Post-Occupancy Evaluation should be planned soon after handing over the building to its users, and six months after inauguration, to obtain satisfaction levels and identify positive and negative aspects of the design and building in use. This phase is essential to close the design process loop and provide important feedback to new projects (HEFCE, 2006).

3. The Local School Design Process

The design process described above is recommended to attain a desired school building quality. In the local Brazilian public school context this process is as yet not fully adopted, as shown in Figure 2.

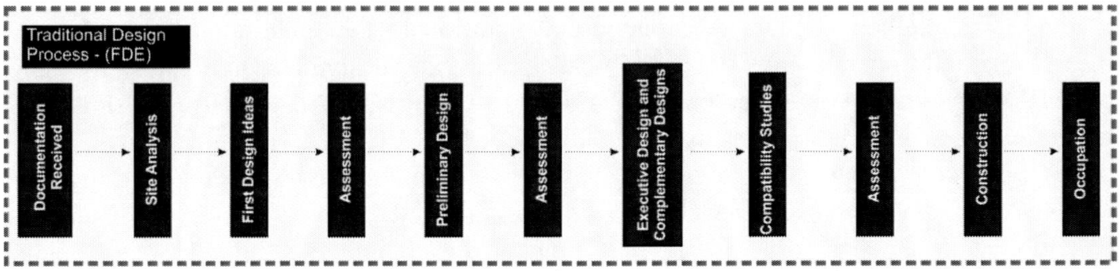

Figure 2. A structure of the local school design process based on the process administered by FDE.

Many efforts have been made to improve the design of school buildings, especially in the State of São Paulo, through the efforts of FDE, including the periodic engagement of well-known professionals (Deliberador, 2010; Ornstein, 2005). The public school design process of the State of São Paulo, managed by FDE, is initiated by contracting local accredited architectural firms. The chosen professionals receive a prefixed brief, elaborated by the State Education Board and construction detailing instructions, based on modular design and prefabricated components, made of concrete. The brief consists in a list of spaces and their dimensions, but the expression of goals, desires or dreams of a school community, are excluded from the local process. The site conditions are provided through specific documentation (topography and foundation condition). In addition, a list of design regulations, that should be consulted, is indicated. A preliminary design is produced with this information, evaluated by FDE and design proceeds, culminating in construction documentation and complementary designs (structural and installations) (FDE, 2000). A special manual developed by FDE should support the treatment of open areas, however, since contracts are poorly remunerated, most designs only indicate very basic landscaping elements.

This process provides few opportunities to introduce new design concepts; especially environmental comfort and sustainability are treated superficially and with no in-depth solution analysis, although the recent adoption of the AQUA Process (FCAV, 2007) is a positive step forward.

Minimum standards of care are only required in the present situation and the process is essentially linear, lacking a feedback loop of positive or negative design solutions. Thus a learning mechanism is excluded. Architectural decision-making deals with often-conflicting parameters, which implies that design methods should display interferences and provide adequate information for the proper conduct in finding the best solution for specific problems.

Also, according to research on local Brazilian professional practice, designers use little reference material while designing, limiting them to codes and eventually some checklists (Graça, 2008). Evaluations in the form of simulations and optimizations are rarely applied to designs and the design process does not include the participation of users. POE studies are not a regular activity of school design teams and building performance assessment data is not readily available for application to new designs.

To obtain an in-depth understanding of the local school design process the results of a case study of structured interviews with professional architects working for FDE are presented. These characterize the specific locally adopted design process (Deliberador, 2010). Figure 3 shows the principle elements present in this process. From these results some opportunities for change can be identified.

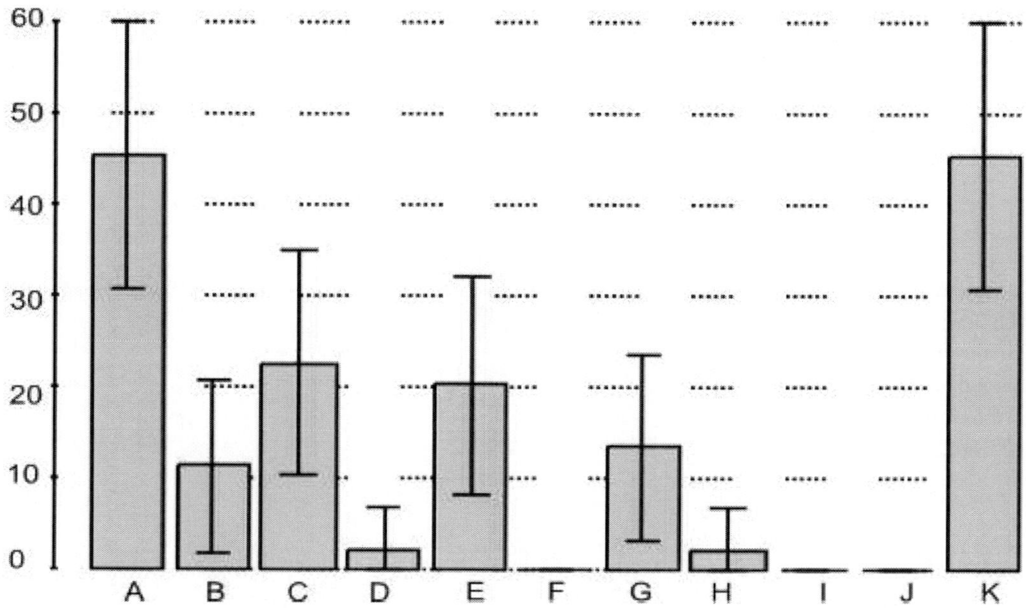

A. Use of Enriched design literature
B. Enriched architectural program
C. Participatory process
D. Multidisciplinary teams
E. Multiple assessment moments
F. Site analysis: risk situations
G. Experts participation in design process
H. Commissioning
I. Post-occupation evaluation
J. Retrofit
Feedback of previous experiences.

Figure 3. Presence in % of characteristics in relation to the elements found in the local school design process.

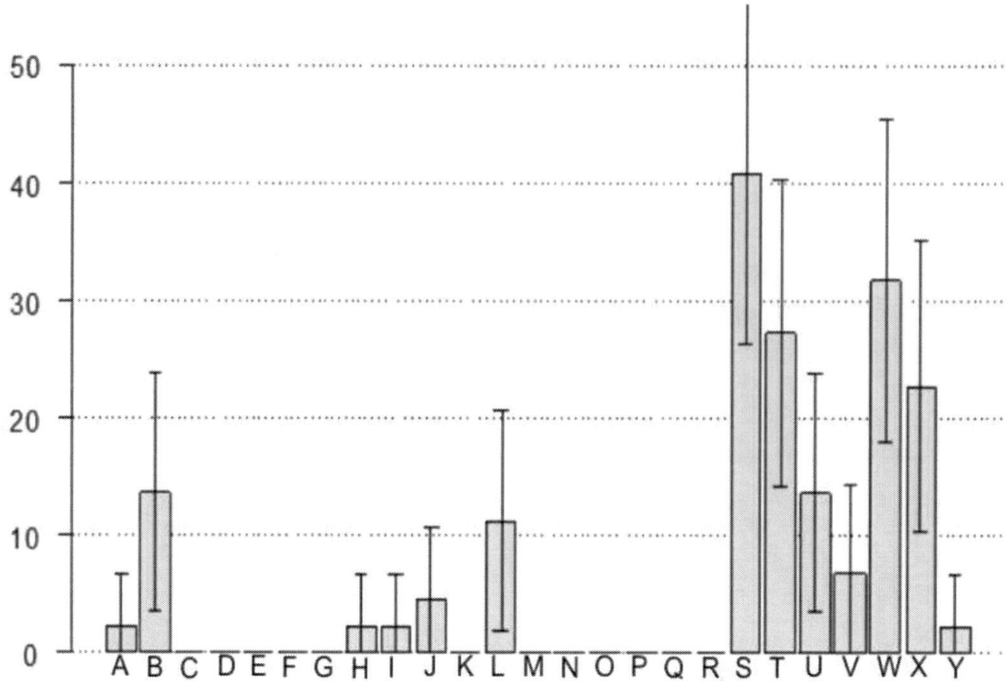

A- Classrooms
B- Welcome Entry
C- Student display space
D- Home base and individual storage
E- Science labs and arts
F- Arts, music and performance
G- Physical fitness
H- Casual eating areas
I- Transparency
J- Interior and exterior vistas
K- Dispersed technology
L- In/outdoor connection
M- Furniture
N- Flexible spaces
O- Campfire Space
P- Watering hole space
Q- Cave space
R- Multiple intelligences
S- Daylight
T- Natural ventilation
U- Learning, lighting / color
V- Sustainable elements
W- Local Signature
X- Connected to the community
Y- Bringing it all together.

Figure 4. Design parameters in % (patterns) mentioned by professionals as criteria in the design process.

The results of these structured interviews of 44 designers, working for FDE, showed that they rarely adopt the specific aspects of the reference design process detailed above (Figure 1). This means that the latest information coming from the literature on school building design is not formally incorporated into new design solutions and the local design process can be considered relatively poor in relation to international recommendations. Justifications for this poverty are the low remuneration for designs and the extreme time limits imposed on the process. Previous experiences are shown to be the guiding principles of the local design process.

The design parameters, or patterns, detailed in Nair and Fielding (2009) also demonstrated a low presence in this design process (Figure 4). A justification may be found in the difficulty in applying the patterns to local conditions. But the result also shows that there is a lack of discussions during the briefing phase. Local public schools are devoid of educationally enriching spaces, such as laboratories, art studios and a resource centre or library, reflected in a rigid brief based essentially on the traditional classroom with fixed dimensions.

It is important to note that this space, the classroom, was also not mentioned as deserving special attention during the design phase, showing little leeway given to discussions and therefore reduced interest by designers to improve the classroom space.

Construction techniques and infrastructure technology was also not mentioned as being an important concern in school building design.

Designers did mention environmental comfort, especially lighting and ventilation as primary design aspects, due in part to attend FDE demands. Thus, classrooms should have cross-ventilation, solved through the provision of large windows on the exterior wall and high clearstory windows to the corridor side of the room. This, in a double-loaded corridor scheme however, does not provide for adequate levels of ventilation in a room with 30 children.

The case study also investigated the design approach of the professionals, and the origin of their first form-giving ideas. This question was asked, in part, to ascertain if sustainability is taken into account at an early stage and if the concept of bioclimatic architecture is used as a basis for preliminary design decisions. Here the answers were not conclusive, since many designers cite sustainability in a very loose sense, saying that their architecture is always concerned with these issues.

However they are unable to detail these questions. As mentioned before, thermal comfort is solved through the introduction of openings on two opposite sides of classrooms, a ceiling slab under the roof cover and if necessary, some solar shading panels for external wall openings with east or west orientation.

North orientations are often not treated, causing problems in tropical latitudes in the southern hemisphere for large parts of the day. Acoustics is a much neglected comfort issue, probably because FDE does not stipulate special attention to this question and designers are restricted in their specifications of acoustic ceilings and barriers, necessary in many locations and spaces of schools.

Recently FDE has obliged designers to justify their choice and detailing of sun protection elements, since many school buildings are shown to have poor conditions, where children sit exposed to full sun early in the morning or late in the afternoons.

Figure 5 shows the results of what designers considered important concepts of environmental comfort. 80% of the designers interviewed mentioned that they are concerned with the detailing of shading devices. Also, natural ventilation and orientation of facades were both mentioned as important aspects, probably because they are FDE requirements. These, no doubt are important issues, but not the only ones. When analyzing the extended list of elements that constitute a quality school environment, the local design process, practiced by architects working for FDE, must be considered poor in relation to the reference process, with room for improvements.

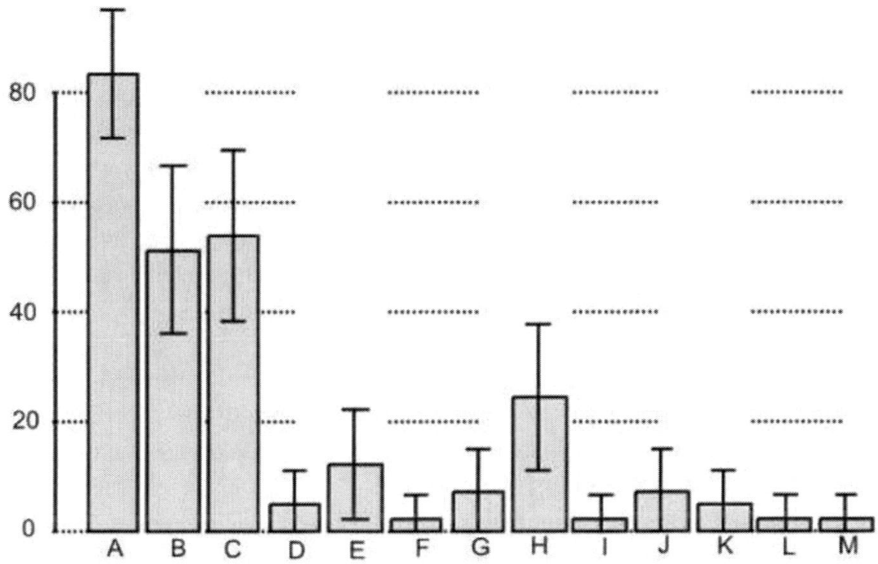

A. Brise Soleil
B. Solar Orientation of building
C. Natural Ventilation
D. Colors
E. Openings - windows
F. Eaves – roof extensions
G. Landscaping aspects
H. Construction Materials
I. Balconies/ verandas
J. FDE specific concepts
K. No concepts applied
L. Openings in the ceiling
M. Furniture orientation.

Figure 5. Inclusion in % of design concepts for thermal and natural lighting comfort.

In relation to sustainability (Figure 6) very few aspects are mentioned, although many respondents mentioned sustainability as their guiding concept. On the other hand, FDE recently adopted new criteria for school design, which include several important questions of sustainability, such as the building's urban relation, energy and water efficiency, waste and maintenance management, environmental comfort and health standards of indoor spaces, as well as construction site conditions.

Architects interviewed feel that the construction system, based on prefabricated concrete elements can be considered sustainable, since it generates less waste. On the other hand, many designers criticized this building system, adopted mainly to reduce costs. It is in need of a re-evaluation in view of the many problems encountered during the construction phase and in finished buildings.

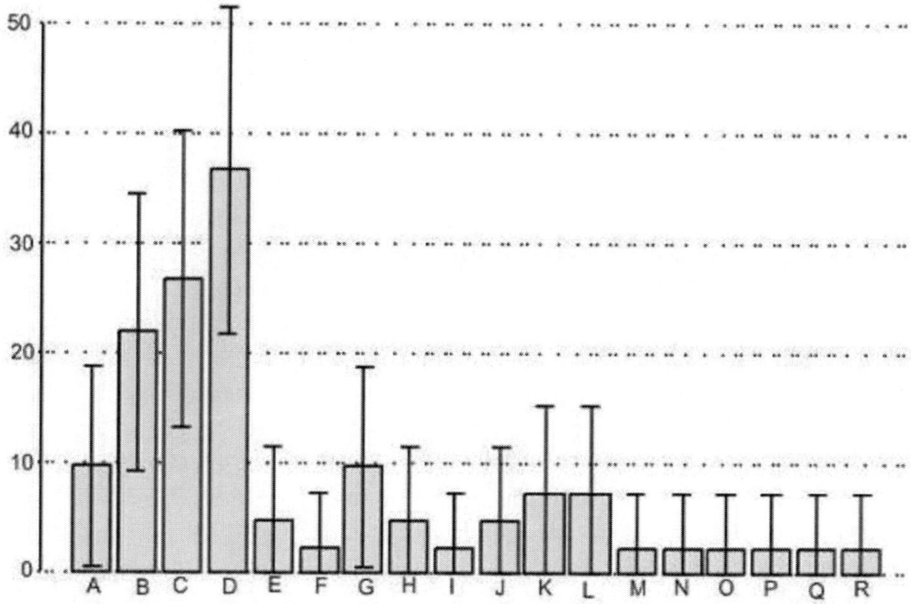

A. Good architecture
B. Building System
C. Ventilation
D. No concepts applied
E. Brise Soleil
F. Urban aspects
G. Site Planning
H. Landscaping aspects
I. Soil Permeability
J. Energy efficiency
K. Materials
L. Water reuse
M. The building in relation to its urban surrounding
N. Social aspects
O. Cut and fill of the site
P. Acoustic aspects
Q. Bicycle stands
R. Waste.

Figure 6. Inclusion in % of design concepts for sustainability.

The survey in relation to sustainability further showed that the majority of local professionals are as yet not prepared to conduct a complete certification process such as LEED (U.S. Green Building Council, 2009) with its wide range of issues to be met.

Thus sustainability is only present in school buildings through the use of energy efficient light fixtures and reduced number of water taps and sanitary installations located only on the ground floor.

4. Functional Aspects of the School Environment

While an appropriate design process can ensure recommended procedures, such as participation of users and experts and the application of evaluation tools, the content of the design discussions and decisions must also be addressed. As demonstrated, most evaluation phases of a recommended design process concentrate their attention on environmental comfort issues. Some POE studies assess user satisfaction (Gomes da Silva et al., 2009; Kowaltowski, 2011; Ornstein, 2005), however functional aspects of school environments are rarely detailed in relation to the provision of appropriate spaces for desired activities in Brazilian studies.

This maybe due to the fact that classrooms are still very much a standard 50 sq. m size and the available area per student is usually the only indicator of a school's functional quality. However, this indicator is a poor measure in relation to school performance, or the possibility of flexible use of space to accommodate a variety of activities (Ornstein et al., 1993; Kowaltowski et al., 2001).

Looking for specific results on the assessment of functional elements in public school buildings in Brazil one finds that the learning environments are less than stimulating, often austere and dull, with rather dark color schemes that were chosen for easy maintenance and not for aesthetic reasons. Service areas, toilets, and locker rooms are the biggest problems in relation to hygiene and maintenance.

Most schools have a lack of spaces to keep cleaning products. Custodial implements and broken furniture and equipment are often kept in inappropriate places, like the back of classrooms, under stairs and at the end of corridors. These conditions give an impression of disorder and reduce functional space for more productive educational activities. Maintenance in general is a problem in the public sector. Often parent associations are called upon the task to repair broken parts or to give the building a general overhaul.

The schoolyard is another place with a lack of visible attention. Most schools have no landscape project and the planting of trees, shrubs and flowerbeds is left to the school staff, with less than ideal results. Playgrounds often do not exit and shaded areas with benches for children to have their snacks are few. The setting of many school buildings is thus arid and less than pleasant. To overcome typical problems of rainwater drainage school administration will pave parts of the outdoor spaces between buildings in a haphazard way, increasing the impression of a lack of professionalism in public education.

Students and staff opinions about their school environment can be conflicting. Students are often neutral in their judgment of functional aspects of a school, but tend to evaluate negatively space available in classrooms, academic resources (books, maps etc.) and school maintenance. Most school administrators complain about the high number of students per class and the corresponding size of classrooms.

Supervision of buildings is considered difficult due to layout problems and the many often-unplanned additions. The academic staff feels the lack of proper laboratories and resource centers affects teaching quality negatively. At times, to alleviate such conditions further adaptations are made, often with poor results (Kowaltowski et al., 2001).

Internationally, school performance is regularly studied in relation to socioeconomic, methodological and pedagogical as well as educational and environmental factors (Hamaty and Lines, 1999; Kowaltowski, 1980, Monteiro et al. 1993; MOORE and Wong, 1997;

Sanoff, 1994, Valiant, 1996,). There are studies that show, in the United States of America for example, that the application of tests like the SAT (Scholastic Aptitude Test) presents conflicting relationships, such as high scores in schools with large sophisticated facilities, equipment and specialized staff, while small schools show a more favorable psychological climate for communication and control of educational activities (Andrews et al. 2002; Duran-Narucki, 2008).

Thus the research linking school performance to environmental factors, such as school size, are not always conclusive. The concept of "behavior settings" was developed by Barker and Gump (1964). Groups of users and their activities, fixed in space and supported by architectural elements were observed. Research measuring the relationship between the wealth of "settings" in schools was developed in the 60's, creating indicators for the quality of school buildings.

Other studies on the optimal size of a school, for example, confirm the need for a minimum of 1.50 m2 per student and a maximum size for groups of from 13 to 20 students per school activity. Schools should also have a limit of 500 students per period (Knatz, 1970; Kowaltowski, 1980).

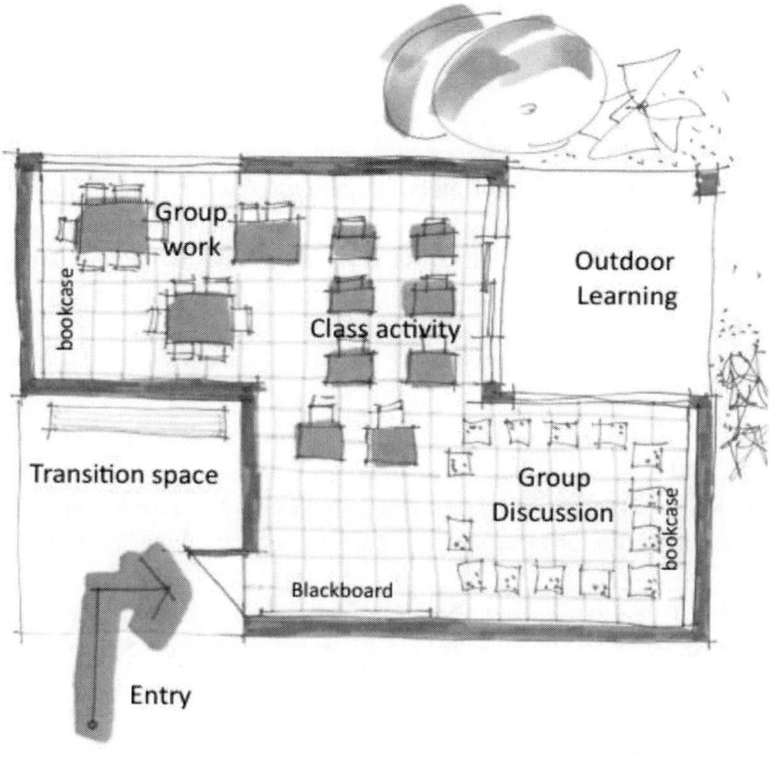

Figure 7. Example of a "Z" shaped classroom configuration based on the idea presented by Nies and Hougsted, 1997.

Other studies concentrate on the shape and size of classrooms (Sanoff, 2001b; Nies and Hougsted, 1997; Lippman, 2004). Various shapes, such as "L" or "Z" shaped classrooms are recommended to enable different activities to take place simultaneously and allow for flexible furniture layouts (Figure 7). In Brazil these discussions have not yet reached the public sector and classrooms are used in a traditional row of desks arrangement for 30 to 45 students with the teacher's desk at the front of the room. This classroom model, in many discussions contains a hidden curriculum of discipline, order and control over students. Silence is encouraged to maintain students focused (Lawn et al., 1999).

The functional discussions of the school environment should be closely related to pedagogical concerns. Specific pedagogies recommend a set of teaching methods and activities to attend to the curriculum content and educational goals. These should also be expressed through architectural language and detailing adopted for a school building design.

Montessori and Waldorf schools have successfully created their own school architecture. Waldorf schools, specially, tend to express specific educational concepts in the shape and building layouts of its architecture, based on the principles of organic design and strongly influenced by the first schools created by the founder of this pedagogy, Rudolf Steiner in the early 20th century (Alvares, 2010).

New ways of designing creative spatial arrangements are more accepted in so-called alternative schools. Some professionals can be shown to have had important positive influences on humanizing the school environment and questioning the "hidden curriculum" in school architecture.

Hertzberger (1986, 2005) is such an architect, who through his many innovative designs, especially for Montessori schools in the Netherlands, has made school buildings less forbidding. Classrooms open to the outdoors and ample corridors permit children to linger and do group or homework. Circulation spaces are particularly valued, giving transparency to the building as a whole. These spaces are then turned into educational streets, rather then mere connecting spaces. A rich array of other educational opportunities is also given through open access to laboratories, a well-equipped and inviting library and pleasant eating-places.

Beyond considerations about the classroom configuration, functional discussions should touch on the variety of spaces needed to provide a rich educational environment to students and staff. As shown in designs by Hertzberger, the library should be a special place and the theories of learning should be the basis for the discussion of school design (Bransford et al., 1999; Valiant, 1996).

From recent studies on the brain we know that each brain is unique, but that it functions better when users work in groups, thus the social ingredient is important. Also there are better cognitive connections with specific activities such as music. We also know that stress affects learning negatively, but emotions (passion for something) can enhance learning (Rashid and Zimring, 2008).

Lackney (1998) has translated this research data into design recommendations based on at least twelve principles of brain-compatible learning that have emerged from research:

1. Uniqueness – every single brain is totally unique.
2. Impact of threat or high stress can alter and impair learning and even kill brain cells.
3. Emotions are critical to learning – they drive our attention, health, learning, meaning and memory.
4. Information is stored and retrieved through multiple memory and neural pathways.
5. All learning is mind-body – movement, foods, attentional cycles, drugs and chemicals all have powerful modulating effects on learning.
6. The brain is a complex and adaptive system – effective change involves the entire complex system.

7. Patterns and programs drive our understanding – intelligence is the ability to elicit and to construct useful patterns.
8. The brain is meaning-driven – meaning is more important to the brain than information.
9. Learning is often rich and non-conscious – we process both parts and wholes simultaneously and are affected a great deal by peripheral influences.
10. The brain develops better in concert with other brains – intelligence is valued in the context of the society in which we live.
11. The brain develops with various stages of readiness.
12. Enrichment – the brain can grow new connections at any age. Complex, challenging experiences with feedback are best. Cognitive skills develop better with music and motor skills.

Lackney (1998) recommends that the school environment should be a stimulating place and design should not be reduced to space allocation type of design. Schools should have an appropriate color scheme and provide places that invite social interactions, such as nooks and alcoves to isolate a small group from the hustle and bustle of a typical school corridor. Both active and passive spaces are important. The brain should be given enough oxygenation through outdoor connections and body movement. Feeling secure is important, as is a variety of form, color and light. Children and teachers must be able to change spaces spontaneously, personalizing rooms and walls and transforming a staircase, for instance, into a small theater. The flexible use of space must be encouraged, which does not mean providing neutral spaces, but environments that demonstrate a variety of possible uses. Access to equipment and learning material should be easy, to make the most of available resources. The community around the school should play an active part in daily school activities, so that learning can be more practical and occur in the real world.

When designing quality schools with function in mind, accessibility can no longer be left to chance and the seven principles of universal design (UD) must be consciously incorporated in all design decisions (ADA, 1905):

1. Equitable Use
2. Flexibility in Use
3. Simple and Intuitive Use
4. Perceptible Information
5. Tolerance for Error
6. Low Physical Effort
7. Size and Space for Approach and Use

In schools, the question of inclusion does not only refer to barrier free environments, but primarily teaching the correct attitudes towards people with disabilities. When the physical environment reflects such attitudes the school community is naturally reminded of its obligations. Inclusion should be achieved through almost seamless design decisions and aesthetic decisions can be free of the dictates of regulations, as long as the design incorporates the spirit of these guidelines and standards.

Brazil has a national norm on accessibility since 1989 (ABNT, 2004; MEC, 1997), which has been revised several times already, demonstrating the dynamics present in regulatory statutes. Most existing school buildings however do not as yet comply with the minimum standard of accessibility. In the State of São Paulo new buildings will attend the 2004 regulation and many older schools have been fitted with elevators to give physical accessibility. Elevators are expensive items and must be constantly maintained to be of optimal use. Often such elevators are locked for special use only and this condition does not represent the spirit of accessibility fully. Physical accessibility in new buildings is mainly solved through the substitution of stairs with ramps, which, depending on the circulation flow of students during class intervals may not always be the best design solution. Ramps take up large spaces and stairs are often more efficient in moving people without disabilities, when strategically located.

The wide variety of disabilities, existing in a population, is still not fully dealt with in the public school environment. Thus, especially concerning the proper educational attention given to deaf or blind children and students with multiple disabilities, schools need better training of staff and innovative teaching methods to deal with the wide variety of problems that may arise from a full inclusion program. Such programs are based on the premise of giving every child the same opportunities, guarantying equal conditions and independence regardless of the disadvantages present. According to Sassaki (1997, 2005) schools must invest in:

- Removal of architectural barriers for the free movement and access to all functional spaces, urban or built, of physically disabled persons, whether using wheelchairs or crutches etc...
- Guaranteeing communication, be it interpersonal, written or virtual
- Revising teaching and work methods to be inclusive
- Making tools and equipment able to be handled by all
- Making public policies free of bias
- Supporting attitudes that fight bias, prejudice, stigmas and stereotyping
- Eradicating discrimination altogether.

The physical environment needs specific adaptations and academic material and equipment must be available to make the educational experience rich for all users. It is not enough to turn a school building into an accessible place. The urban context is often the mayor hurdle for children with disabilities to be included into normal public education. Knowing that some 15% of Brazilians have some kind of disability, barriers must be removed in homes, on streets and for public transport (IBGE, 2012).

Not only attitudes need to be taught, but also, especially in developing countries, like Brazil, many life-skills should be taught in school and for this to happen the traditional bare classroom is not the best place. Children need to become passionate about good nutrition, learn how to cook, do gardening, get notions on house construction techniques and repairs, as well as acquire interests in manual tasks or crafts. Playing a musical instrument is a further life enriching activity, as important as keeping fit, through physical activities and a healthy diet. Such activities do not need sophisticated environments, an example of a redesigned cafeteria is shown in Figure 8.

Many activities can happen in multipurpose rooms with some provisions such as a few sinks, sturdy workbenches and a small industrial kitchen. Music can be taught in many places, but needs isolation from other activities at times and spaces for performances, where a sufficiently large welcoming entrance hall can suffice. Physical activities can be enhanced through such activities as taking care of a school vegetable garden and orchard.

To make school users feel able and safe, other factors from environmental psychology should be considered such as: personal space, privacy, territoriality and wayfinding or orientability (Gifford, 2001). Personal space is essential to give users their identity. People feel uncomfortable when others encroach upon their personal space and become defensive, aggressive or recluse.

Thus space per person, as defined for the classroom to be a minimum of 1.5 sq. m, should not only be reserved for functional purposes, but be sufficient, therefore a little more generous, to avoid the feeling of crowding in schools.

While space per se is important, the arrangements or layouts of spaces affect human behavior as well. Looking constantly at the back at a colleague's neck is considered negative and can cause stress in children. Sociopetal arrangements are encouraged for discussion groups, but sociofugal nooks are also needed, to from time to time, isolate oneself from the group, do some thinking or reading alone and to be able to relax (Lawson, 2001).

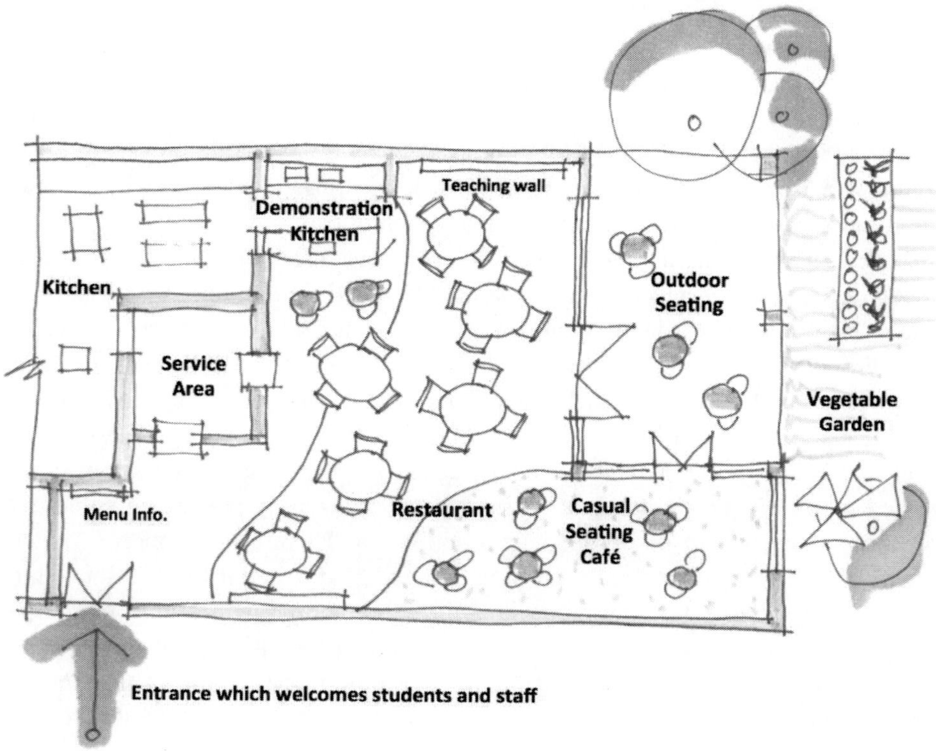

Figure 8. Example of floor plan of a school cafeteria that incorporates life skills such as cooking and nutrition classes as well as casual eating with contact to the outdoors, based on the design proposal presented by Gorman et al., 2007.

For these reasons it is important to provide a variety of arrangements to allow many positive learning settings to happen and support flexible personal and social interactions in the classroom and in the more public spaces of schools, in corridors and in the schoolyard.

When children start school the new environment can be overwhelming and confusing. This may cause stress and influence negatively the child's outlook on learning. Thus, the school building should have a well-organized plan that can be "read" through its volumetric configuration from the street. Entrances should be marked clearly and be positioned in strategically and visible locations. Keeping corridors open enhances wayfinding. Thus double loaded corridor schemes should be avoided. Giving a free view of public indoor spaces and the movement of people through the building is an important factor to consider in school design.

A clear color scheme, distinctive forms in different parts of buildings and using specific marks, such as: totems, sculptures, fountains with water sounds, along the circulation spaces may help, as well, to organize wayfinding. Views to the outside, to get ones bearings, are also recommended. Of course, good signage and clear maps, displayed in key places, are essential in large buildings.

The feeling of belonging is increased through the concepts of personal space and territoriality. Children should have their own place for a period of time and the territory of a class should be personalized by a group of users for the term or school year. Privacy must be assured in specific situations of stress. Children also need places to be on their own for a while, rest and have some privacy. A gradient of public to private spaces should be available within the classroom, which can even offer some nooks for an occasional necessary nap. Thus a teacher/parent conference needs an isolated place with acoustic barriers. Also each school needs spaces to isolate a child with an acute problem, be it emotional or physical.

Feeling safe is closely related to knowing ones way around, thus wayfinding has to do with a feeling of being at home, of belonging and being accepted by the school community. Of course, school environments must ensure the physical safety of its users first. Thus, high places need railings, ramps and stairs must be detailed according to regulations, exits must be located and dimensioned for fire and emergency egress and sharp and protruding edges must be beyond the reach of people walking. Hot surfaces from furnaces or ovens and dangerous equipments in general must be in locked or supervised areas. Swimming pools need fences and constant supervision when in use.

Some spatial configurations, although safe according to regulations, can cause anxieties. Dark narrow corridors, dead-end plan configurations and untidy places are some examples and such conditions should be avoided in schools. Vandalism occurs more in such areas (Kowaltowski, 1980). But it should be remembered that vandalism is often simply the damage to school property through faulty detailing, thus school architecture must be robust, but not austere to the extreme, giving an impression of being almost a prison. Again, transparency or the visibility of functional spaces is important to avoid vandalism and even bullying, a continuing major problem in many schools. From data of occurrences, criminal acts are mainly located in non-supervised spaces, such as wash and locker rooms and in the public areas of schools. Although questions of privacy do not recommend transparency in rest rooms, their access should be located where the movement of people, circulation of students and staff, provides natural and constant supervision.

From psychology there is good evidence that behavior is the result of many components: the specific psychology of an individual user; his or her socio-economic situation and cultural values, expressed in a way-of-life; the age and development level of that person and the perception of a place and situation (Fisher, 2000; Walden, 2009). Children pass through distinct phases and a school environment must provide the proper conditions for positive components of that growth to thrive. Motor coordination is to be fostered at each stage of development, by giving opportunities to test movement precision, equilibrium and speed. Sensorial stimulation can occur through architectural elements, color and light. Anthropometry is another issue in school building design. Children come in all sizes and grow often in rather rapid bouts during one school year. School furniture must be adjustable and flexible to accommodate these variations with efficiency and comfort (Cardon et al., 2004).

The influence a school building has on user behavior is not based on simple formulas (Tanner, 2000; Maxwell, 2000; Higgins, et al., 2005). Human perception is impacted by a variety of factors, not least the human climate established through staff and student attitudes. However from positive examples the learning environment can learn as well. Autonomy should be given to students. This will improve their self-esteem and learning willingness. Their artwork should be displayed (Killeen et al., 2003). Innovations are important to stimulate interest and can be extended to new ways of teaching and sporadic changes introduced into the environment (Lackney and Jacobs, 2002; Lang, 2002; Nigaglioni, 2005). Experimentation is important and should be encouraged, including changing the use of spaces. Students and staff should also have a say in proposing changes, for instance, in relation to a new color scheme for the school.

Designers should remember that their professional training impacts their perception of space and their understanding of architectural necessities. This different perspective is positive, but needs clarification with users in participatory design processes (Pennartz and Elsinga, 1990). Discussions of the differences in perception are excellent educational opportunities as well. Teachers in particular need training to understand the full potential of a building or the classroom design (Bissell, 2002; Pauly, 1991).

Technically speaking, there are still some controversies over the effect of levels of lighting and room proportions, relative scale of architectural elements and color, although much is known on the minimum standards necessary for a productive learning environment in relation to air quality, temperature, ventilation, and noise in classrooms. Other factors affect pupil's perception of space, such as proportions, seating arrangements and as mentioned previously questions of security. However it is difficult to pinpoint specific elements. What can be made clear is that the environment must create the perception of a well organized, clean and aesthetically pleasing place (Earthman, 2009).

Students must feel comfortable, have the support of furniture that is both attractive and allows for flexibility in postures during class time. This time should be used efficiently and not be wasted with lengthy adjustments of the room and its equipment. Equipment must be plentiful and belong to a group of learners, so it will be explored to its fullest. Valuing student work must be encouraged through exhibitions that are renewed from time to time and not left do abandonment on old billboards. Once the school has created its positive outlook on its own users it will be able to extent a positive image to the wider community (Rittelmeier, 1994). This can then be drawn into the school to enrich learning experiences of students, teach civility, handle antisocial behavior (bullying, vandalism, drug use, gang activities and even extreme shyness) and reinforce positive behavior (discipline, good achievement rates, altruism, voluntarism, entrepreneurship, creativity, etc.) (Sanoff, 1992; 2002).

5. Developing Tools to Stimulate Better School Design

To obtain a positive learning environment in schools periodic evaluations are necessary to assess if the building, its teaching methods and administration attend the educational goals set by the local school community and society at large. These assessments need technical evaluations with equipment to measure levels of environmental comfort and through observations to check the adequacy of space, learning material and their use by staff and students. Quantitative and qualitative evaluations are important. Thus, subjective assessments should be documented to get a sense of the users' feelings about the place, beyond neutral satisfaction rates.

This type of evaluations can be of importance for renovations and new project. Technical and subjective data should be made available to the school community to judge the need for change and embrace a new design process to improve the learning environment. This process needs to follow reference processes as closely as possible. From the characterization of the local, Brazilian school design process, opportunities for change became evident (Deliberador, 2010). Thus, the introduction of many evaluation phases (simulations, calculations, debates) was considered positive by local professionals. Commissioning and POE studies were seen as possible positive new actions to improve the local design process. Professionals are already obliged to justify specific design elements such as *brise soleil* and will, in the future, need to include a sustainability assessment according to the AQUA process.

The participation of users, in the decision making process, is still a major hurdle in improving the local design process. Public schools have to follow specific patterns of standards and be of equal quality for all communities. Thus questions of ethics come into play.

Giving users a voice, but not the power to decide, is a difficult issue, which should be discussed and overcome. On the other hand, a participatory process is also an educational opportunity and should be at least conducted in that spirit. Making users aware of their school environment, its potential is a way of stimulating wider debates on education and focus, not only on pedagogy, but also on the building, the schoolyard and the urban area surrounding a public school.

Woolner (2009) investigated the "pros" and "cons" of participation and concluded that wide consultation will increase the reconciliation of conflicting views and may complicate and lengthen a design process. However the same study argues that such tensions and difficulties are worth encountering, especially if a well-organized collaborative design process is adopted. It is important to warn against pro-forma participation, going merely through the motions, without making people feel truly involved. True participation depends on an ongoing respectful and genuine dialogue, involving a wide range of people and ideas.

To support such user participation debates some tools can be applied to stimulate reflection and focus discussions on specific school building design issues. Adapting the DQI for schools for local Brazilian conditions is an important task, since the use of this tool forces the design team to reflect on a number of essential issues (Gann, et al., 2003).

Spaces to exhibit student work work and stage special events.

Areas for students to work in groups in an informal way and be able to study, as well as permit spontaneous interactions.

Informal eating places outside the common cafeteria space.

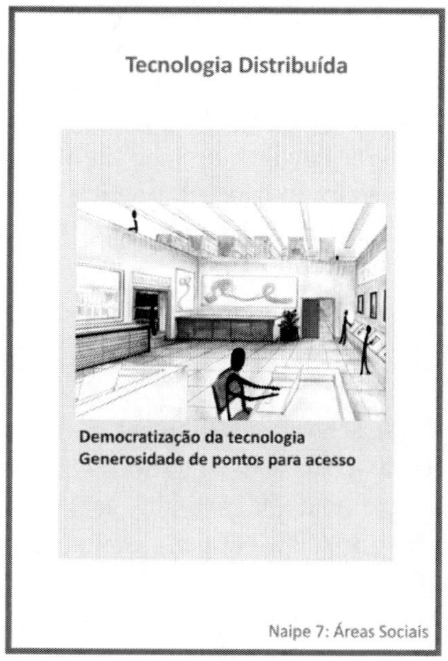

Distributed technology throughout the school building and grounds, to permit free access and spontaneous interaction.

Figure 9. Example of a set of playing cards of the school design process game showing essential social spaces that should be provided.

Other methods can also enrich the pre-design process. The authors developed a school design card game, as shown in Figure 9 (A, B, C and D). This game consists of a deck of cards organized in sets of suits. A similar set of cards was already developed and used in social housing design by the authors with some promising results (Kowaltowski and Granja, 2011).

The game is divided into 12 sets of suits, each with four options or issues to be discussed. These are: different pedagogies; the classroom; the schoolyard; the school in its urban context; environmental comfort aspects; special spaces; socializing areas; important design concepts; zoning issues; administrative support areas; student support areas and circulation spaces.

48 different cards were developed to represent design aspects to be discussed among design process participants. The aim of the use of such a game is to stimulate discussions. Also, the cards can be used to rate the degrees of importance for each aspect of design represented on the cards and select specific issues, concepts or types of spatial arrangements as priorities.

The graphic representation of each aspect of school design is important to focus on similar understanding of a design issue. Participants can discuss the issues and give priorities to aspects, suggest other ways of solving problems and add design examples to be explored. Just as the "patterns" of Nair and Fielding (2009), the school design card game does not offer specific solutions, but opens the process to possibilities and can enrich the search for solutions.

Other tools can be applied and developed to enrich the school design process. The design team should be supported with good access to design information, case study reports and easy to use evaluation tools with their application manuals. This can be in the form of sites, such as examples found in the UK and the USA. CABE now under the UK Design Council (Commission for Architecture and the Built Environment) and "Design Share" come to mind. Easy access and up-to-date information are essential for such support to be of real value. Thus a local Brazilian school design support organization is envisioned.

Conclusion

A quality school environment depends on a multitude of factors such as: the school community, its goals and aspirations and its available resources, in the form of staff, teaching equipment and material and the school building and grounds. Functional considerations should always be related to pedagogical questions and the social and cultural conditions of the setting of the school.

Since schools should serve the future, reflections on the dynamics of education must accompany school design processes. The philosopher Edgar Morin teaches us lessons that should be applied to our schools of the future (Morin, 2001). These lessons are mainly concerned with the uncertainty of the world and its knowledge, as well as the necessity to think locally in a globalized world. But foremost is the lesson on humanity for schools of the future, and this should apply to the design of school buildings as well. Schools should be places that are welcoming, inspirational and beautiful, connected to nature, have comfortable, healthy and safe conditions, with a caring atmosphere that embraces the inclusion of multiple intelligences.

The design process to attain such a school environment must have as its basis some fundamental concepts of: humanization, generosity and flexibility, sustainability and accessibility for all. The question of a quality environment does no longer concern the quality of specific building components but a wider range of concerns and their interconnections.

The quality of school designs under local Brazilian and especially in the State of São Paulo conditions is based on specific processes, directed by FDE the organizing and administrative organ for school design and maintenance in the State. In such a design process the particular experience and abilities of designers are of extreme importance. From investigations on this process it is clear that conditions of environmental comfort are rarely considered as the starting point for a design and a variety of design alternatives to solve comfort problems are rarely generated and evaluated. The quality of the design is thus left to chance, since the process lacks sufficient structure. Important evaluation phases and an early design debate are missing with an imposed program or brief. When comparing the local design process to some of the recommendations found in the literature on high performance school buildings, opportunities for change can be identified, such as awareness of risk situations, presented by difficult sites, that demand special attention. Professional designers also show an interest in getting involved in a richer design process. Finally, the characterization of the local school design process indicates the need for better assessment tools and access to information organized and directly applicable to local school design issues.

REFERENCES

ABNT- Associação Brasileira de Normas Técnicas, 2004. Acessibilidade de pessoas portadoras de deficiência a edificações, espaço, mobiliário e equipamento urbanos, NBR-9050.

ADA-Adaptive Environments Center, Inc., B.F.E., Inc., 1995. The Americans with Disabilities Act Checklist for Readily Achievable Barrier Removal, The Americans with Disabilities Act.

Aldrich, R.S., 2011. A whole systems approach. *Libr. J.* 136, 30–33.

Alexander, C., Ishikawa, S., Silverstein, M., 1977. A Pattern Language: Towns, Buildings, Construction. Oxford University Press.

Alvares, S., 2010. Traduzindo em formas a pedagogia Waldorf (Master Dissertation).

Andrews, M., Duncombe, W., Yinger, J., 2002. Revisiting economies of size in American education: are we any closer to a consensus? *Econ. Educ. Rev.* 21, 245–262.

Barker, R.G., Gump, P.V., 1964. Big School, Small School: High School Size and Student Behavior. *Stanford Univ. Pr.*

Benedikt, M.L., 2008. Human Needs and How Architecture Addresses Them. University of Texas Press, Austin, TX, USA.

Bissell, J.M., 2002. Teachers'construction and use of space.

Boman, E., Enmarker, I., 2004. Factors affecting pupils' noise annoyance in schools: The building and testing of models. *Environ. Behav.* 36, 207–228.

Bordass, W., Leaman, A., Eley, J., others, 2006. A guide to feedback and post-occupancy evaluation. Usable Building Trust.

Bransford, J., 2000. How people learn: Brain, mind, experience, and school. National Academies Press.

Cardon, G., De Clercq, D., De Bourdeaudhuij, I., Breithecker, D., 2004. Sitting habits in elementary schoolchildren: a traditional versus a "Moving school". *Patient Education and Counseling.* 54, 133–142.

Chase, J., 1996. Blueprint for a Green School. Scholastic Trade.

CHPS, 2002. Best Practice Manual. The Collaborative for High Performance Schools, Calif., USA.

Crandell, C.C., Flexer, C., Smaldino, J.J., 2004. Sound Field Amplification: Applications to Speech Perception and Classroom Acoustics, 2nd ed. Singular.

Deliberador, M.S., 2010. O processo de projeto de arquitetura escolar no Estado de São Paulo: caracterização e possibilidades de intervenção (Masters Dissertation).

Dudek, M., 2008. Schools and Kindergartens: A Design Manual, 1st ed. Birkhäuser Architecture.

Duran-Narucki, V., 2008. School building condition, school attendance, and academic achievement in New York City public schools: A mediation model. *J. Environ. Psychol.* 28, 278–286.

Earthman, G.I., 2009. Planning Educational Facilities: What Educators Need to Know, 3rd ed. RandL Education.

FCAV, F.C.A.V., 2007. Referencial Técnico De Certificação, Edifícios do setor de serviços – Processo AQUA, Escritórios - Edifícios escolares. São Paulo, SP.

FDE, 2000. Normas-de-apresentação-projetos-FDE (Norma).

Figueiredo, F.G., 2009. Processo de Projeto Integrado para melhoria de desempenho ambiental de edificações: dois estudos de caso (Masters Dissertation).

Fisher, K., 2001. Building Better Outcomes: The impact of school infrastructure on student outcomes and behaviour. Department of Education, Training and Youth Affairs (Australia).

Ford, A., 2007. Designing the Sustainable School. Images Publishing Dist. Ac.

Ford, A., Hutton, P., 2007. A Sense of Entry: Designing the Welcoming School. Images Publishing Dist. Ac.

Gann, D., Salter, A., Whyte, J., 2003. Design Quality Indicator as a tool for thinking. *Building Research and Information.* 31, 318–333.

Gifford, R., 2001. Environmental Psychology: Principles and Practice, 3rd ed. Optimal Books.

Gomes da Silva, V., Kowaltowski, D.C.C.K., Pereira, P.P.R., Minette, L., Deliberador, M.S., 2009. Analysis of the Building Process and of Energy Efficiency of Public Schools in the State of São Paulo, Brazil, in: International Conference: Sustainable Architecture and Urban Development. Presented at the CSAAR 2009, Tripoli, Lybia, pp. 255–270.

Gorman, N., Lackney, J.A., Rollings, K., Huang, T.T.-K., 2007. Designer schools: The role of school space and architecture in obesity prevention. *Obesity.* 15, 2521–2530.

Graça, V.A.C. da, 2008. A integração dos aspectos de conforto ambiental no projeto de escolas: uso da metodologia axiomática e de exemplos simplificados (Doctoral Thesis).

Graca, V.A.C., Cornelie Knatz Kowaltowski, D.C., Diego Petreche, J.R., 2007. An evaluation method for school building design at the preliminary phase with optimisation of aspects of environmental comfort for the school system of the State Sao Paulo in Brazil. *Build. Environ.* 42, 984–999.

Hamaty, G., Lines, D., 1999. Planning for Schools of the Future: A building program is a golden opportunity to restructure schools to better meet the needs of students and educational programs. School Planning and Management.

HEFCE, H.E.F.C. for E., 2006. Guide to post occupancy evaluation. Higher Education Funding Council for England.

Hershberger, R.G., 1999. Architectural Programming and Predesign Manager. McGraw-Hill Professional Publishing, New York, USA.

Hertzberger, H., 1986. Herman Hertzberger: Recent works. Stichting Wonen of Amsterdam.

Hertzberger, H., 2005. Herman Hertzberger Lessons for Students in Architecture. 010.

Heschong Mahone Group, 1999. Daylighting in Schools: An Investigation Into Relationship Between Daylighting and Human Performance; Detailed Report. Pacific Gas and Electric Company.

Heschong, L., 2003. Windows and Classrooms: a Study of Student Performance and Indoor Environment: Technical Report, Public Interest Energy Research Program (PIER) and New Buildings Institute, Inc and Heschong Mahone Group. California Energy Commission.

Higgins, S., Hall, E., Wall, K., Woolner, P., McCaughey, C., 2005. The impact of school environments: A literature review. Design Council.

Holtz, M.J., 2004. Phase 4: Construction – Commissioning, in: Assessing Building Performance. Butterworth-Heinemann, pp. 62–71.

IBGE, 2012. Censo Demográfico e Contagem da População (Banco de Dados Agregados). Brasília.

ISO, 1994. Performance standards in building -- Checklist for briefing: Contents of brief building design (No. 9699). Standard ISO.

Killeen, J., Evans, G., Danko, S., 2003. The role of permanent student artwork in students sense of ownership in an elementary school. *Environ. Behav.* 35, 250–263.

Knatz, D.C.C., 1970. High School Building Design In Relation To New And Changing Teaching Methods And Their Goals (Masters Dissertation).

Kowaltowski, D., 2011. Arquitetura escolar: o projeto do ambiente de ensino, 1. ed. Oficina de Textos, São Paulo, SP.

Kowaltowski, D., Borges Filho, F., Labaki, L., Ruschel, R.C., Bertoli, S.R., Pina, S.A.M.G., 2001. Melhoria do Conforto Ambiental em Edificações Escolares na Região de Campinas, SP (Relatório científico FAPESP). Campinas, SP.

Kowaltowski, D.C.C.K., 1980. Humanization In Architecture: Analysis Of Themes Through Highschool Building Problems.

Kowaltowski, D.C.C.K., Granja, A.D., 2011. The concept of desired value as a stimulus for change in social housing in Brazil RID D-5953-2011. *Habitat Int.* 35, 435–446.

Lackney, J., 1998. 12 Design Principles Based on Brain-Based Learning Research. Design Share.

Lackney, J., 2006. Designing Healthy Schools Our Children Deserve. DesignShare. 4937 Morgan Avenue South, Minneapolis, MN 55409-2251. Tel: 612-929-6129; Tel: 612-929-3520; Web site: http://www.designshare.com.

Lackney, J.A., 2000. Educational Facilities: The Impact and Role of the Physical Environment of the School on Teaching, Learning and Educational Outcomes. Univ. of Wisconsin Milwaukee.

Lackney, J.A., 2001. The State of Post-Occupancy Evaluation in the Practice of Educational Design, in: Of EDRA 32. Presented at the International Conferece of the Environmental Research Association, Edinburgh, UK.

Lackney, J.A., Jacobs, P.J., 2002. Teachers as Placemakers: Investigating Teachers' Use of the Physical Learning Environment in Instructional Design. For full text: http://schoolstudio.engr.wisc.edu.

Lang, D.C., 2002. Teacher Interactions within the Physical Environment: How Teachers Alter Their Space and/or Routines Because of Classroom Character. (Doctoral Thesis).

Larsson, N., 2009. The integrated design process: theory, history, demonstration. International Initiative for a Sustainable Built Environment - iiSBSE.

Lawn, M., Grosvenor, I., Rousmaniere, K., 1999. Silences and Images. Peter Lang Publishing.

Lawson, B., 2001. Language of Space, 1st ed. Architectural Press.

Lippman, P.C., 2004. The L-Shaped Classroom: A Pattern for Promoting Learning. DesignShare. 4937 Morgan Avenue South, Minneapolis, MN 55409-2251. Tel: 612-929-6129; Tel: 612-929-3520; Web site: http://www.designshare.com.

Maslow, A., 1943. A theory of human motivation. *Psychol. Rev.* 50, 370–396.

Maxwell, L., 2000. A safe and welcoming school: What students, teachers, and parents think. *J. Archit. Plan. Res.* 17, 271–282.

Maxwell, L.E., Chmielewski, E.J., 2008. Environmental personalization and elementary school children's self-esteem. *J. Environ. Psychol.* 28, 143–153.

MEC, M. da E. e do D., 1997. Portadores de Deficiências Físicas. Acessibilidade e utilização dos equipamentos escolares (Cardernos Técnicos No. I.). Brasília.

Monteiro, C., Loureiro, C., Roazzi, A., 1993. A satisfação como critério de avaliação do ambiente construído: um estudo aplicado ao prédio escolar, in: Proceedings of ENTAC. Presented at the Encontro Nacional de Tecnologia do Ambiente Construído, São Paulo, SP, pp. 873–884.

Moore, R.C., Wong, H.H., 1997. Natural Learning: The Life History of an Environmental Schoolyard. Mig Communications.

Moreira, D. de C., Kowaltowski, D.C.C.K., 2009. Discussão sobre a Importância do Programa de Necessidades para a Qualidade no Processo de Projeto em Arquitetura. *Revista ANTAC, Ambiente Construído.* 9, 31–45.

Morin, E., 2001. Seven Complex Lessons in Education for the Future. Unesco.

Myhrvold, A., Olsen, E., Lauridsen, O., 1996. Indoor environment in schools–pupils health and performance in regard to CO2 concentrations, in: Indoor Air. Presented at the Indoor Air, The Seventh International conference on Indoor Air Quality and Climate, Nagoya, Japan, p. 369–71.

Nair, P., Fielding, R., 2009. The Language of School Design: Design Patterns for 21st Century Schools Fully Revised 2nd Edition, 2nd ed. Designshare, Inc.

Nies, J., Hougsted, S., 1997. Z-Shaped Classroom Supports Technology, Enhances Learning. *School planning and management.* 36, 34–36.

Nigaglioni, I., 2005. Reinventing the Traditional Classroom. *Educational Facility Planner.* 40, 3–8.

Ornstein, S., 1997. Postoccupancy evaluation performed in elementary and high schools of Greater Sao Paulo, Brazil - The occupants and the quality of the school environment. *Environ. Behav*. 29, 236–263.

Ornstein, S., Neto, J.B., da Silva Mileo, A.P., Martins, C.A., 1993. O desempenho dos edifícios da rede estadual de ensino: o caso da grande São Paulo: avaliação técnica, primeiros resultados. Faculdade de Arquitetura e Urbanismo da Universidade de São Paulo, Departamento de Tecnologia, São Paulo, SP.

Ornstein, S.W., 2005. Post-Occupancy Evaluation in Brazil, in: Anais: OECD/PEB Experts' Group Meetings on Evaluating Quality in Educational Facilities. Presented at the OECD/PEB Experts' Group Meetings on Evaluating Quality in Educational Facilities, Paris, France, pp. 135–143.

Pauly, E., 1992. The Classroom Crucible: What Really Works, What Doesn't, And Why. Basic Books.

Peña, W.M., Parshall, S.A., 2001. Problem Seeking: An Architectural Programming Primer, 4th ed. Wiley.

Pennartz, P., Elsinga, M., 1990. Adults, Adolescents, And Architects - Differences In Perception Of The Urban-Environment. *Environ. Behav*. 22, 675–714.

Rashid, M., Zimring, C., 2008. A review of the empirical literature on the relationships between indoor environment and stress in health care and office settings - Problems and prospects of sharing evidence. *Environ. Behav*. 40, 151–190.

Rittelmeyer, C., 1994. Schulbauten positiv gestalten: wie Schüler Farben und Formen erleben. Bauverlag.

Samad, Z.A., Macmillan, S., 2005. The Valuation Of Intangibles: Explored Through Primary School Design. Presented at the Proceedings of the CIB W096 conference on —Designing value: new directions in architectural management, Lyngby, Denmark, pp. 1–8.

Sanoff, H., 1992. Integrating Programming, Evaluation and Participation in Design: A Theory Z Approach. Avebury.

Sanoff, H., 1994. School Design, 1st ed. Wiley.

Sanoff, H., 2001a. A Visioning Process for Designing Responsive Schools, NCEF National Clearinghouse for Educational Facilities. ed. Washington DC.

Sanoff, H., 2001b. School building assessment methods. National Clearinghouse for Educational Facilities U.S. Dept. of Education, Office of Educational Research and Improvement, Educational Resources Information Center.

Sanoff, H., 2002. Schools Designed with Community Participation. (Educational Facilities). NCEF, Washington DC.

Sassaki, R.K., 1997. Inclusão: construindo uma sociedade para todos. WVA Editora e Distribuidora Ltda, Rio De Janeiro, RJ.

Sassaki, R.K., 2005. Conceitos de acessibilidade.

Schneider, M., 2002. Do school facilities affect academic outcomes? National Clearinghouse for Educational Facilities. Accessed 12-05-07 at: http://www. edfacilities. org/pubs/outcomes. pdf.

SPATE, 2004. Sistema de Planejamento e administração de tempo e espaço em edifícios educacionais.

Tanner, C.K., 2000. The influence of school architecture on academic achievement. *Journal of Educational Administration*. 38, 309–330.

U.S. Green Building Council (U.S.G.B.), 2009. LEED 2009 for Schools New Construction and Major Renovations Rating System. U.S. Green Building Council, Washington, DC.

Upitis, R., 2004. School architecture and complexity. *Complicity: An International Journal of Complexity and Education.* 1, 19–38.

Valiant, B., 1996. Turn on the Lights! Using What We Know about the Brain and Learning To Design Learning Environments. (ERIC Document Reproduction Service No. ED460568), Issue Trak: A CEFPI Brief on Educational Facility Issues.

van der Voordt, T.J.M., Vrielink, D., van Wegen, H.B.R., 1997. Comparative floorplan-analysis in programming and architectural design. *Design Studies.* 18, 67–88.

Walden, R., 2008. Schools for the Future, 1st ed. Hogrefe Publishing.

Watson, C., 2003. *Review of building quality using post occupancy evaluation* (PEB Exchange Programme on Educational Buildings). Paris, France.

Wong, F.W.H., Lam, P.T.I., Chan, E.H.W., 2009. Optimising design objectives using the Balanced Scorecard approach. *Design Stud.* 30, 369–392.

Woolner, P., 2009. *Building Schools for the Future through a participatory design process: exploring the issues and investigating ways forward, in: BERA2009.* Presented at the The British Education Association Conference, Manchester, UK, pp. 1–17.

Zimring, C., Rashid, M., Kampschroer, K., 2005. *Facility Performance Evaluation* (FPE), Whole Building Design Guide.

In: Noise and Ergonomics in the Workplace
Editor: Paulo Henrique Trombetta Zannin

ISBN: 978-1-62417-573-2

Chapter 2

NOISE IN THE SCHOOL ENVIRONMENT

Paulo Henrique Trombetta Zannin [*]
and Carmen Lúcia Pezzette Loro
Laboratory of Environmental and Industrial Acoustics and Acoustic Comfort,
Federal University of Paraná, Brazil

ABSTRACT

The acoustic quality of a standard classroom in a public school was evaluated based on measurements of the reverberation time and ambient noise level inside and outside classrooms. The reverberation time was measured in three conditions: a) unoccupied classroom, b) classroom occupied to 50% of its capacity, and c) classroom occupied 100%. Concomitantly to the assessment of the acoustic quality of classrooms, teachers and students' perception of noise inside and outside the school was also investigated. Teachers and students stated that they perceive the presence of noise in their daily school activities and that it is bothersome to a greater or lesser degree. Among their main reactions to noise, the teachers pointed out: 1) the need to speak more loudly, 2) loss of concentration, 3) irritability, restlessness and edginess, and 3) headaches, fatigue and buzzing in the ears. The main reactions among the students were: 1) difficulty to concentrate, 2) headache, 3) the need to speak more loudly, 4) edginess, 5) buzzing in the ears, 6) irritability and restlessness, and 7) fatigue.

1. INTRODUCTION

Education plays a fundamental role in the formation of modern society. The importance of education for humans is expressed thus by renowned Brazilian educator Paulo Freire: *"The fountainhead of man's hope is the same as that of his educability: the incompleteness of his being of which he has become aware. It would be a sorry contradiction if, incomplete and aware of this incompleteness, man were not engaged in a permanent process of hopeful search. This process is education.*" (Freire, 2000).

---

[*] E-mail: paulo.zannin@pesquisador.cnpq.br, paulo.zannin@gmail.com.

Education is as essential in today's societies as it was in the past. The most formal education takes place in classrooms, where learning involves intensive verbal communication between teachers and students (Lubman and Sutherland, 2003). However diverse the techniques for knowledge transmission aided by multimedia resources, which have become increasingly common, nothing replaces the teacher-student relationship which is basically developed in the classroom. It is therefore crucial for classrooms to provide the conditions required for the satisfactory development of teaching and learning activities, especially teacher-student communication. It is in this context that the importance of classroom acoustics stands out (Zannin et al., 2012).

Hagen et al. (2002) stated that: "High levels of noise in the classroom cause students to become more easily tired, interfering in a harmful way in their academic performance." Classrooms with good acoustics aid learning, making it more efficient and pleasurable (Lubman and Sutherland, 2003). Maxwell and Evans (2000) showed that the exposure of preschool children to noise is linked to difficulties in learning to read and write.

Upon examining the effects of chronic aircraft, rail and road traffic noise on school children's cognitive performance, Stansfeld (2002) observed the following reactions of children exposed to environmental noise: 1) deficits in sustained attention and visual attention; 2) difficulties in concentrating for noise-exposed children compared to children from quieter schools, according to teacher's reports; 3) poorer auditory discrimination and speech perception; 4) memory impairment for tasks that require high processing demands; and 5) poorer reading ability and school performance on national standardized tests.

Frangos (2003) reports that, according to the National Center for Educational Statistics (USA), teachers miss about two days per year due to vocal fatigue, caused by raising their voices to talk over the ambient noise in and around the school classroom. Seep et al. (2002) reports that in 1998, 7.9 billion dollars were spent on school buildings in the United States, and wonders why acoustic problems in classrooms are endemic when the solutions are not necessarily expensive. According to Seep et al. (2002), "the main reason is not lack of resources but lack of perception of the problem."

Martins (2002) considers that today's children are forced to learn in noisy classrooms, where the noise both inside and outside the classroom impairs their ability to hear their teacher's voice, thus requiring high levels of attention. According to Martins (2002), noise can negatively affect both teachers and students.

Santos (1993) considers the following steps necessary in the assessment and control of noise in schools: 1) identification of the internal and external sources of noise and measurement of their sound pressure levels; 2) identification of the constructive model; 3) construction of the school (location and orientation); 4) distribution of its internal spaces and allocation of activities inside the school by sectors of activities; 5) definition if its constructive elements for good sound insulation: windows, doors, walls and floors; and 6) specific acoustic treatment for special areas. Classrooms can be treated acoustically based on measurements and/or simulations of parameters such as reverberation time, speech transmission index, definition, etc. (Zannin et al. 2009, Zannin et al. 2011, Zannin et al. 2012).

The main scope of this chapter is to study the indoor and outdoor acoustic quality in public schools. The classrooms evaluated here were built according to a modular design, which determines that they should have the following characteristics: all the classrooms have the same dimensions, constructive details, structural design, and electrical design.

As can be seen from the above, the quality of a school environment depends on several variables. Therefore, the objective of this chapter is to evaluate the acoustic quality of the school environment using both an objective approach – involving measurements of acoustic parameters such as reverberation time and ambient noise levels – and a subjective approach, involving an investigation of teachers and students' perception of noise inside and outside these classrooms based on the application of questionnaires.

2. Description of the Classrooms

The Department of Education of the state of Paraná has developed the following classroom designs: 1) Design Standard 023; 2) Double Floor Design Standard 023; and 3) Design Standard 022 (Fundepar, 1990).

Among these three classroom models, the one most widely used is Design Standard 023, and is therefore the focus of study in this chapter (see Figures 1 and 2).

Figure 1 shows the façades of two schools in different regions of the state of Paraná, about 600 km apart: 1) in the metropolitan area of the state capital, Curitiba, and 2) in the municipality of Medianeira, western Paraná.

Figure 1. Left: Photograph of the façade of a school located in metropolitan Curitiba. Right: Photograph of the façade of a school located in the municipality of Medianeira, in western Paraná.

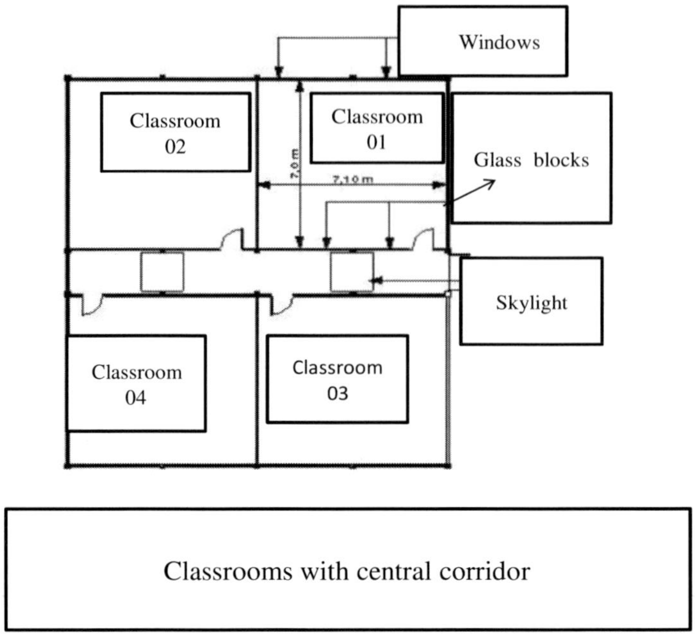

Figure 2. Schematic diagram of classrooms, and central corridor with ceiling skylight and Glass blocks embedded in the walls separating it from the classrooms.

Figure 2 indicate that the classroom design consists of two classrooms separated by a central corridor. This pattern can be repeated over and over. All the public schools built according to this design have the same characteristics, i.e., the same architectural, structural and electrical designs. Therefore, these classrooms are built to the same dimensions, with the same construction details and finishing materials.

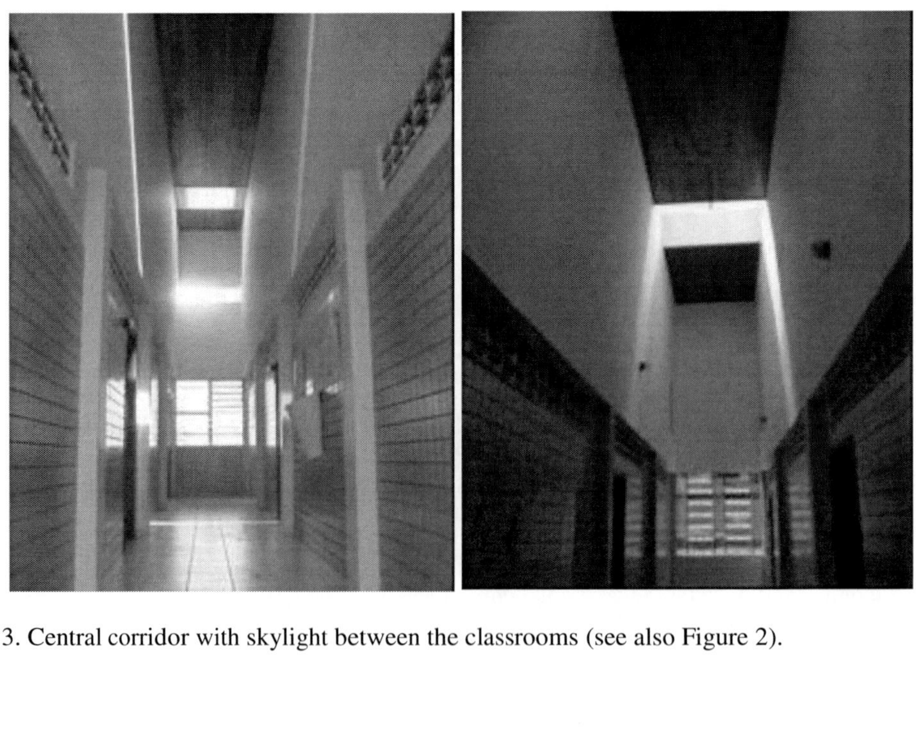

Figure 3. Central corridor with skylight between the classrooms (see also Figure 2).

Figure 4. Detail of the glass blocks embedded in the wall separating the classrooms from the corridor, and glass ventilation shutters for air circulation and additional lighting in the classrooms (see also Figure 2).

The design specifies reinforced concrete construction (columns, beams and ceiling), plastered brick walls, ceramic floor tiles, molding on the main illumination opening, glass with iron ventilation and plywood door. Figure 2 illustrates the classroom's dimensions: width 7.0 m; length 7.1 m and height 3.1 m. Each classroom has a seating capacity of 40 students.

The central corridor is 6 meters high (Figure 3), and each module has a polycarbonate skylight and ventilation shutters. The upper part of the walls between the corridor and the classrooms has ventilation slits and permanent illumination provided by glass blocks (see Figure 4).

3. Ambient Noise Levels – Indoors and Outdoors– in Schoolrooms

Ambient noise (indoor and outdoor) is one of the parameters that affect the acoustic quality of classrooms (Karabiber and Vallet, 2003; Astolfi and Corrado, 2003; Zannin and Zwirtes, 2009).

Several countries such as Australia, Brazil, France, Germany, Greece, Italy, Japan, New Zealand, Portugal, Spain, Turkey, the United Kingdom and the United States recommend limiting indoor noise to a given acceptable level. The highest level varies according to the type and uses of rooms, as shown in Table 1 (furnished and unoccupied rooms).

The Brazilian Standard for Acoustic Comfort in Buildings – NBR 10152 (ABNT, 1987) establishes limits for ambient noise in several types of rooms, as indicated in Table 1. For classrooms, for example, this standard defines 40 dB(A) as the limit for acoustic comfort, and 50 dB(A) as the acceptable limit for the room's purpose. In other words, if the ambient noise level is higher than 40 dB(A) but lower than 50 dB(A), the classroom is considered acoustically uncomfortable, but can still be used for its proposed purpose.

Table 1. Indoor noise limits for different room purposes, according to the standards and regulations of various countries (adapted from Karabyber and Vallet, 2003)

Country	Indoor noise descriptor Equivalent sound pressure level dB(A)	Classroom	Library	Music classroom	Halls and corridors	Dining room, gymnasium
Australia	L_{eq}	30 – 35	40 – 45	40 - 45	45 -50	45 – 55
Brazil	L_{eq}	40 – 50	35 – 45	35 - 45	45 -55	45 – 60
France	L_{eq}	38	33	---	---	43
Germany	L_{eq} L_{Max}	30 - 40 40 – 50	30 – 40 40 – 50	---	---	---
Italy	L_{eq}	36	---	40	40	---
Japan	L_{eq}	40 – 45	35 - 40	35 - 40	---	---
New Zealand	L_{eq}	30 – 35	40 - 45	40 - 45	45 – 50	45 – 55
Spain	L_{eq}	40 – 40	35 – 35	---	50 – 50	---
Turkey	L_{eq}	45	---	---	---	60
UK	$L_{eq,30min}$	30 – 35	35	30	45	45
USA	L_{eq}	35* 40**	35* 40**	35* 40**	45	35* 40**

*Maximum one hour-average A-weighed steady background noise level.

**The limits for A-weighed steady background noise levels shall be increased by 5 dB when the noisiest hour is dominated by transportation noise (unsteady background noise).

Values for ambient noise in the USA refer to rooms with a maximum volume of 566 m^3.

Outdoor noise measurements around the school were taken according to criteria established by the Brazilian ABNT – NBR 10151 (2000) standard, which establishes how to perform noise assessments in populated areas to ensure the comfort of the community. The ABNT NBR 10151 standard establishes guidelines to be followed during outdoor measurements, as indicated in Figure 5. Measurements of sound pressure levels inside buildings should be taken at a height of approximately 1.2 m from the floor and at least 2 m away from any reflecting surfaces such as walls, etc.

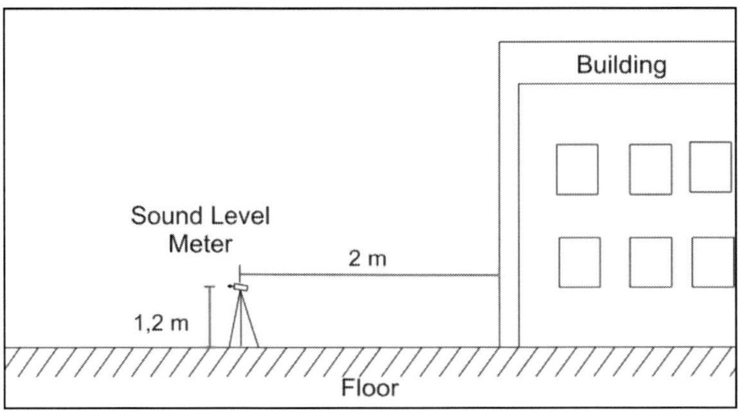

Figure 5. Outdoor noise measurements taken around the school building.

Table 2. Noise levels as a function of the type of land use, according to the NBR 10151

Type of Area	Equivalent sound pressure level, L_{eq} dB(A)	
	Diurnal	Nocturnal
Rural	40	35
Hospitals and Schools	50	45
Homes	55	50
Commercial uses	60	55
Industrial uses	70	60

Table 2 lists outdoor noise levels as a function of the type of land use and time of day, according to the NBR 10151 standard.

4. Reverberation Times for Classrooms

Another important parameter affecting the acoustical quality of classroom is reverberation time – RT. A good definition for reverberation and reverberation time is given by Harris (1994): *"Suppose a source emits sound in a room. The sound, which travels from the source to a listener by a direct air path, without any reflections, is called the direct sound. It is the first sound to reach the listener because it travels from the source directly to the listener by the shortest possible path. It is then followed by reflections from the walls, ceiling, and floor, and from objects within the room. In general, these reflected waves arrive in such rapid succession that they are not heard as distinct repetitions of the original sound. This is because the ear is unable to resolve two sounds if they are spaced closer in time than about 0.06 second. As a result, the series of reflections are heard as a prolongation of sound from the source, after it has stopped. The prolongation of sound after the source has stopped is called reverberation. The reverberation time (RT) of a room is the time required for the sound level in the room to decrease 60 dB after the source is stopped."*

The reverberation time is strongly dependent on: 1) the room volume; 2) the sound frequency in the room, and 3) the total sound absorption in the room. It is important that the reverberation time is suitable for the intended use of the room.

With regard to classrooms, various acoustic standards present reference values for the RT that should be observed in the design of classrooms. Brazil has a national standard for room acoustics, the NBR 12179 (1992), which establishes criteria for the acoustic treatment of enclosed spaces. The establishment of optimal RT values is determined according to the type of use and the volume of the room for the frequency of 500 Hz. The standard that specifies RTs for enclosed spaces (furnished and unoccupied) does not specifically mention classrooms, but presents the curve for the optimal reverberation time at a frequency of 500 Hz for conference rooms, which can be used to evaluate classrooms in view of the latter's similar use for oral communication.

Germany, Japan, the U.K., the U.S. and France have specific technical standards for evaluating the RT of classrooms. In Japan (Fukuchi and Ueno 2004), RT values represent the average in 2-octave bands including 500 Hz and 1000 Hz, and RT is measured in the furnished and unoccupied classroom. In the U.K., the Building Bulletin BB-93 (2003) indicates that RT is given in terms of the maximum mid-frequency reverberation time, T_{mf}, and the average RT in 500 Hz, 1000 Hz and 2000 Hz octave bands, and that RT should be measured in the unfurnished and unoccupied classroom.

In the U.S. (ANSI S12.60, 2002), RT is given as the maximum RT for mid-band frequencies of 500 Hz, 1000 Hz and 2000 Hz, and RT is measured in the furnished and unoccupied classroom. In Germany, the DIN 18041 (2004) standard establishes that RT values represent the average in 2-octave bands including 500 Hz and 1000 Hz, and RT is measured in the furnished and occupied classroom.

Table 3. Recommended reverberation time for classrooms

Country	Reverberation Time, RT [s]	Volume, V in [m^3]
Brazil	RT = 0.5 - for f = 500 Hz RT = 0.6 - for f = 500 Hz RT = 0.7 - for f = 500 Hz	V = 120 V = 285 V = 630
France	$0.4 < RT \leq 0.8$ $0.6 < RT \leq 1.2$	$V \leq 250$ $V > 250$
Germany	RT = 0.5 RT = 0.6 RT = 0.7 RT = 0.8 For furnished and occupied classrooms	 V = 125 V = 250 V = 500 V = 750
Japan	RT = 0.6 RT = 0.7	V ~ 200 V ~ 300
U.K.	$RT = T_{mf}^{+} < 0.6$ $RT = T_{mf}^{++} < 0.8$ $^{+}$Primary schools; $^{++}$Secondary schools For unfurnished and unoccupied classrooms	- -
U.S.	RT = 0.6 RT = 0.7	$V \leq 283$ $283 < V \leq 566$
World Health Organization WHO	RT = 0.6	-

The DIN 18041 also specifies that, in general, the RT of an unoccupied classroom should not be more than 0.2 s above the required value listed in Table 3.

In France, the RT is calculated as the arithmetic mean of the values of RT at the frequencies of 500 Hz, 1000 Hz and 2000 Hz (WHO, 2001). Table 3 lists the RT recommended in different countries according to the volume of the classroom.

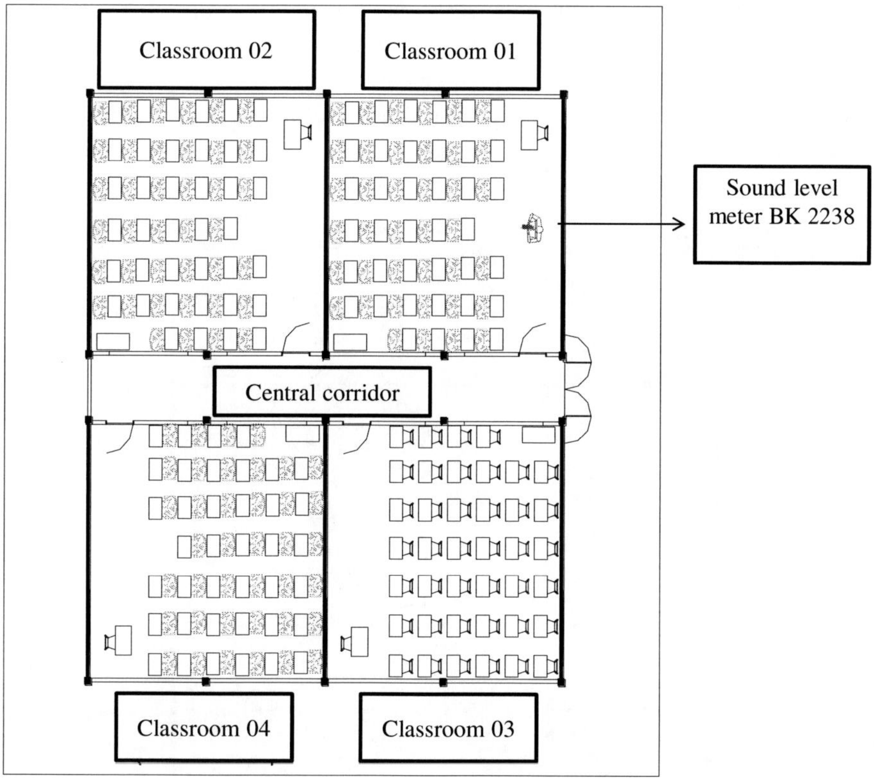

Figure 6. Measurement of ambient noise in classroom 01 with 40 students; Classrooms 02 and 04 with normal classes; Classroom 03 unoccupied. Prevailing situation in Classroom 01: children working individually, some movement and conversation.

5. Evaluation of Ambient Noise and Reverberation Time in Classrooms

The sound pressure levels were measured outside and inside the classrooms for 15 minutes. The ambient noise in the classrooms (closed doors and open windows) was measured in the following conditions:

1. Measurement in classroom 01 occupied by 40 12-year-old students. Classrooms 02 and 04 occupied and classroom 03 unoccupied. Prevailing situation in classroom 01: children working individually, some movement and conversation (see Figure 6).
2. Measurement in classroom 01 occupied by 40 12-year-old students. Classrooms 02 and 04 occupied and classroom 03 unoccupied. Prevailing situation in classroom 1: children working in groups, some movement and conversation.
3. Measurement in classroom 03, unoccupied. Classrooms 01 with normal occupation, children working individually, some movement and conversation. Classrooms 02 and 04 with children working in silence.
4. Measurement in classroom 03, unoccupied. Classrooms 01, 02 and 04 with normal classes. Classrooms occupied by 40 12-year-old students.

5. Measurement in classroom 02, children in silence. Classroom 03 unoccupied. Classrooms 01 and 04 in silence (see Figure 7).
6. Measurement of ambient noise in unoccupied corridor. Classroom 03 unoccupied. Classroom 01, individual tasks; Classroom 02, children working individually, some movement and conversation; Classroom 04, children working in groups, some movement and conversation.

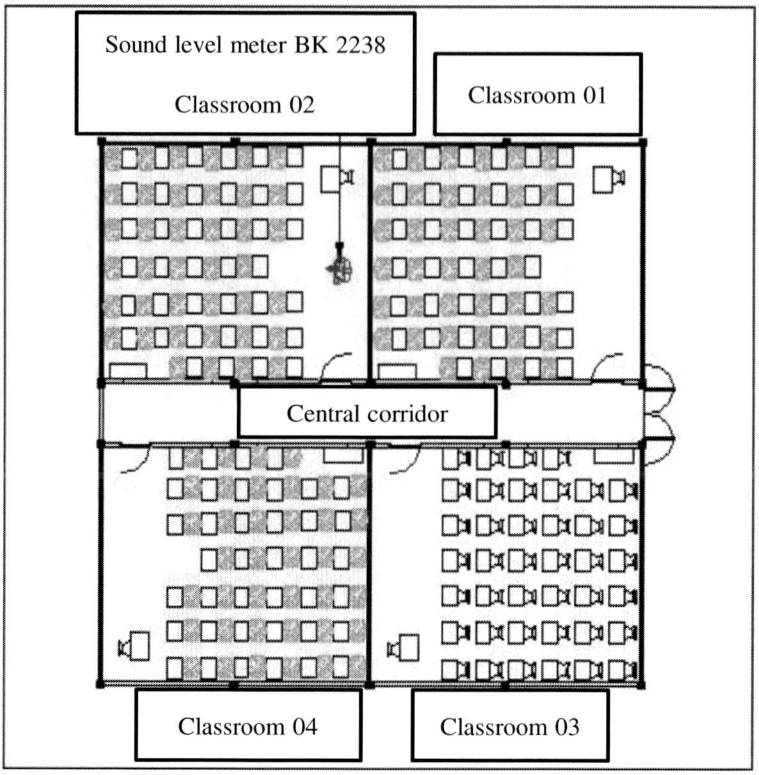

Figure 7. Measurement in classroom 02, in silence; Classrooms 01 and 04 in silence; Classrooms with 40 students; and Classroom 03 unoccupied.

Reverberation time was measured in the following conditions:

1. Furnished but unoccupied classroom
2. Furnished and occupied classroom (see Figure 8)
3. Corridor between two rows of classrooms (see Figure 3)

According to the ISO 3382-1 (2009) and ISO 3382-2 (2008) standards, the RT can be measured by the interrupted noise method and by the integrated impulse response method. Measuring the RT by the interrupted noise method, as described in this chapter, consists of exciting the room with a pseudo-random pink noise and calculating the RT from the room's response to this excitation (Passero and Zannin, 2010).

A common setup to measure the RT by this method comprises: 1) an omnidirectional sound source, 2) a sound power amplifier, 3) a noise generator, 4) omnidirectional microphones, and 5) a sound decay recorder and analyzer (Zannin et al., 2011; Zannin et al., 2012).

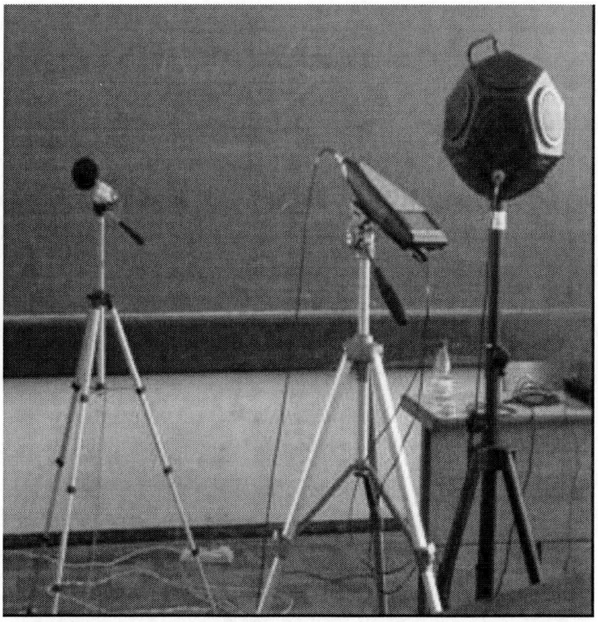

Figure 8. The figure shows the equipment employed to measure the RT – from left to right: microphone, dual-channel real-time sound analyzer and dodecahedron loudspeaker.

The RT described in this chapter was measured with a dual-channel Brüel and Kjaer BK 2260 real-time sound analyzer, a BK 2716 power amplifier, and a BK 4296 dodecahedron loudspeaker. The sound thus generated was captured by a microphone connected to the BK 2260 analyzer, which calculated the reverberation time for each frequency of the spectrum of interest.

These measurements were then transferred to a computer using Brüel and Kjaer BK 7830 Qualifier software, which calculated the mean reverberation time of each classroom (Zannin et al., 2011; Zannin et al., 2012).

6. Teachers and Students' Perception of Noise in School and Classroom

To evaluate teachers and students' perception of noise in the school and classrooms, a questionnaire was created for each group.

The interviews involved 15 teachers and 185 6th and 8th grade students. The purpose of the questionnaire was to investigate the sources of noise in classrooms and the degree of disruption it caused for teachers. The questionnaire for the teachers included a scale of scores to be attributed to each answer: (0) No disruption, (1) Very minor disruption, (2) Little disruption, (3) Moderate disruption, (4) Considerable disruption, (5) Intense disruption, (6) Extreme disruption.

The figures in section 7, which correspond to the interviews, show the mean scores attributed to each answer, i.e., the sum of the scores for each answer was divided by the number of interviewees.

7. Results and Discussion

The school is located in the metropolitan area of Curitiba, in the town of Pinhais, which has no specific legislation for noise emissions. Brazilian cities that do not have their own noise emission regulations must adopt the values established by the NBR 10151 standard described in Table 2. The mean equivalent sound level measured outside the classrooms was L_{eq} = 59.2 dB(A), thus exceeding the level recommended for educational environments, which is L_{eq}= 50 dB(A). Hence, it can be considered that the school is located in a noisy urban area.

To determine the influence of ambient noise passing from one classroom to another, Classroom 3 was measured in the unoccupied but furnished condition, while Classrooms 1, 2 and 4 were occupied to their maximum capacity of 40 students. The mean equivalent sound level measured in Classroom 3, with Classrooms 2 and 4 in complete silence and Classroom 1 with normal classroom noise, was L_{eq} = 56.2 dB(A). In another simulation, Classroom 3 was unoccupied while Classrooms 1, 2 and 4 had normal classes. The mean equivalent sound level measured in Classroom 3 was L_{eq} = 63.3 dB(A). The doors of all classrooms were closed and the windows were left open.

Another set of measurements was taken in the following conditions: the students of Classroom 1 remained under activities, answering a questionnaire with the teacher giving instructions, and under the presence of some movements in the classroom. The mean equivalent sound level measured in Classroom 1 was L_{eq} = 75.2 dB(A) (see Figure 6). Classrooms 2 and 4 had normal classes and Classroom 3 was unoccupied. Another situation analyzed was with Classrooms 1, 2 and 4 with normal classes and Classroom 3 unoccupied. In this situation, the measured ambient noise in classroom 1 was L_{eq} = 80.5 dB(A).

Another test involved measuring Classroom 2 in silence (see Figure 7), with Classrooms 1 and 4 also in silence and Classroom 3 unoccupied. In this condition, the mean equivalent sound level measured in Classroom 2 was L_{eq} = 57 dB(A).

Table 1 lists the limit values of noise levels inside classrooms, which should be measured according to the conditions specified by each national standard. For example: 1) In Japan (Fukuchi and Ueno, 2004), the noise levels are the values measured in a furnished and unoccupied room, with closed doors and windows and HVAC systems turned on. 2) In the U.K., Building Bulletin 93 (2003) proposes that indoor ambient noise levels should be the sum of noise from all sources outside the school, including, for instance, the noise from the ventilation system. The ambient noise level should not include noise from other activities within the school and the room should be assumed to be furnished but unoccupied. 3) In the U.S., the ANSI S12.60 (2002) standard states that the ambient noise level is the sound in a furnished, unoccupied space, including sounds from outdoors, building services and utilities operating at their maximum levels. According to the American standard, the measurement should exclude sound generated by people within the building or sound generated by temporary or permanent instructional equipment. The Brazilian NBR 10152 standard does not impose restrictions to the presence of people or teaching activities within the school building. The only assumption is that the room should be unoccupied and furnished. Therefore, the results obtained in this study cannot be compared with those of standards in which the presence of activities in the school is not allowed during these measurements.

From the above, and considering only the Brazilian Technical Standard 10152 for the assessment of indoor sound environment of classrooms, the results indicate that Classroom 2 (occupied with completely silent students, and the students in the other classrooms as described above) does not meet the national standard, since the measured sound level, L_{eq} = 57 dB(A), exceeds the specified limits. Since this study was based on classrooms that all share the same dimensions and specifications, it can be concluded that none of the classrooms evaluated here meet the national standard for indoor noise limits. Therefore, according to the Brazilian standard, they are uncomfortable rooms.

As can be seen in Figure 9, the teachers were asked to identify the sources of noise that cause the greatest disruption of classroom activities. They cited noise from neighboring classrooms, at level 4, as the main source of disruption, and the voices of teachers in neighboring classrooms, at level 3, as the second source of disruption. The noise generated in their own classroom and noises generated by neighbors around the school were given scores of level 2 and level 1.7 disruption. The latter score, i.e., the fact that neighbors around the school are an important source of disruption, is corroborated by the measurements of sound pressure levels taken outside the classrooms, L_{eq} = 59.2 dB(A), which indicate that the school is located in a noisy urban region, since this value exceeds the limit of 50 dB(A) for school areas, according to the NBR 10151 standard described in Table 2.

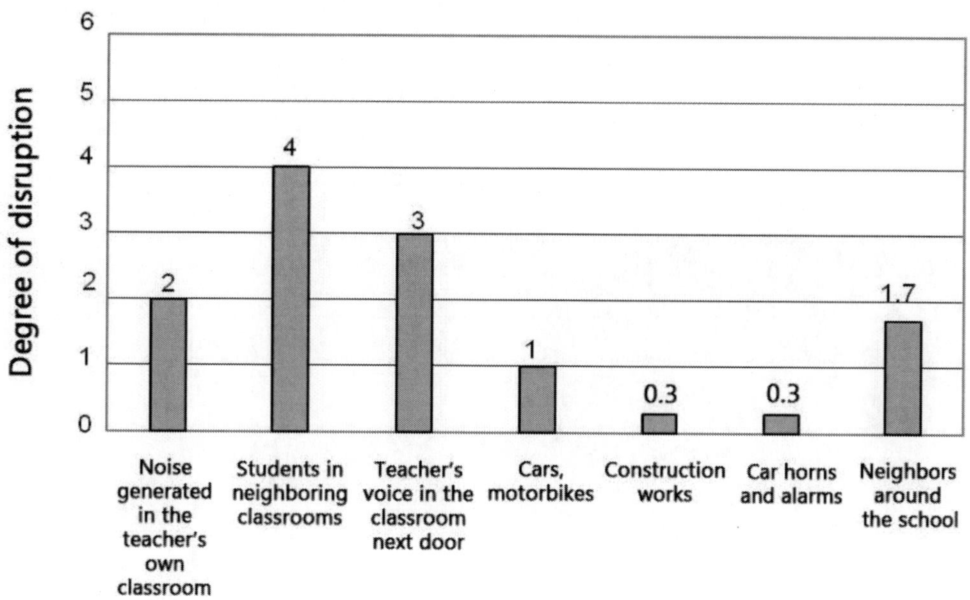

Figure 9. Answers to the question: "What sources of noise cause the greatest disruption of classroom activities?"

Figure 10 indicates that the majority of students, i.e., 70%, consider their classroom noisy. None of the students interviewed, i.e., 0%, considered their classroom silent. In addition to the students' perception of noise in the school environment, 89% of the interviewed students consider that noise in the school environment can impair their learning performance.

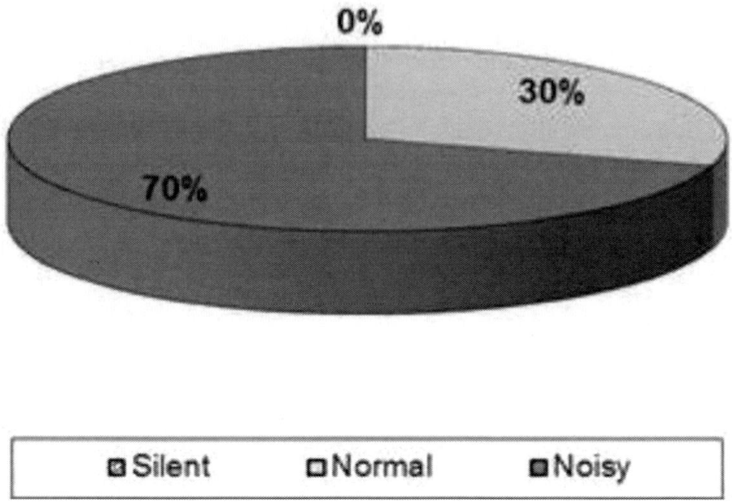

Figure 10. Response to the question posed to the students: "In terms of noise, how would you rate your classroom?"

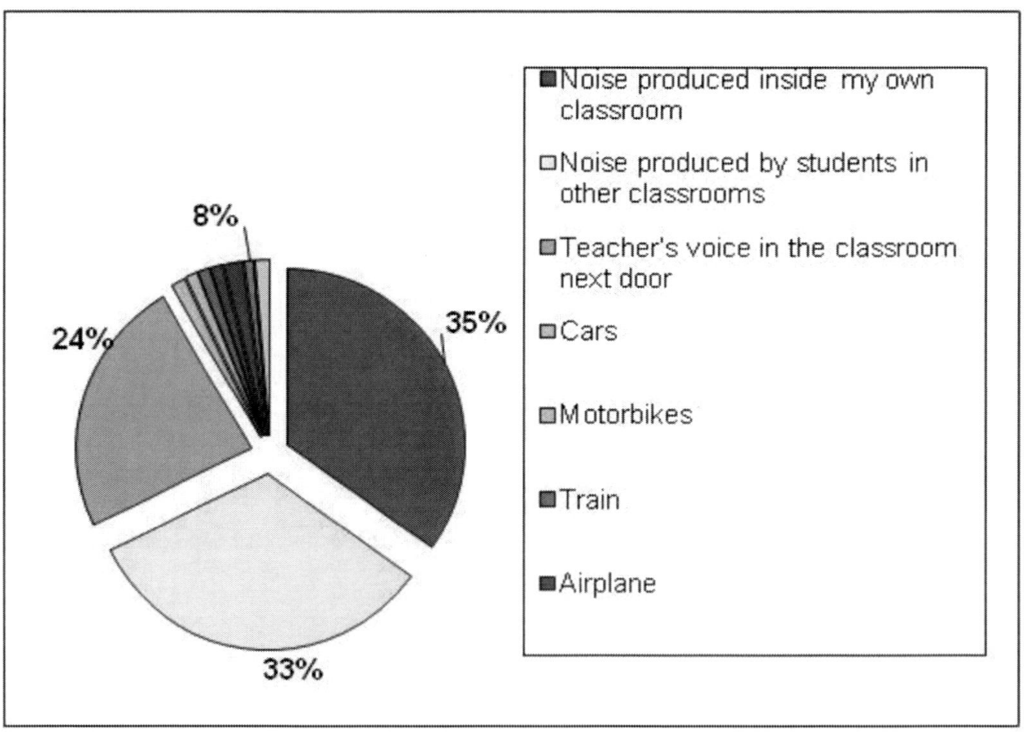

Figure 11. Response to the question posed to the students: "What sources of noise cause the greatest disruption in the classroom?"

As Figure 11 indicates, the students cited as the main sources of disruption in the classroom: 1) noise inside the classroom itself, 2) noise caused by students in other classrooms, and 3) the teacher's voice in the classroom next door. These three sources together represent 92% of the disruption perceived by the students.

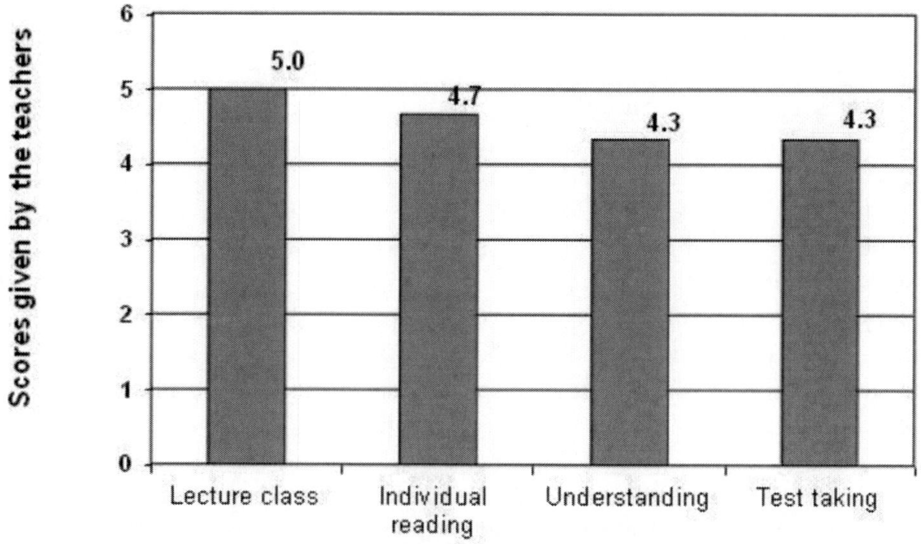

Figure 12. Response to the question: "Which classroom activity is the most affected by noises generated inside the school?"

The teachers indicated, conform Figure 12, that the classroom activities most strongly affected by noise are: lecture classes, individual reading, understanding and test taking. As Figure 12 indicates, the teachers attributed practically the same score – level 4 (considerable disruption) and level 5 (intense disruption) – for all these activities, demonstrating that they are strongly affected by noise.

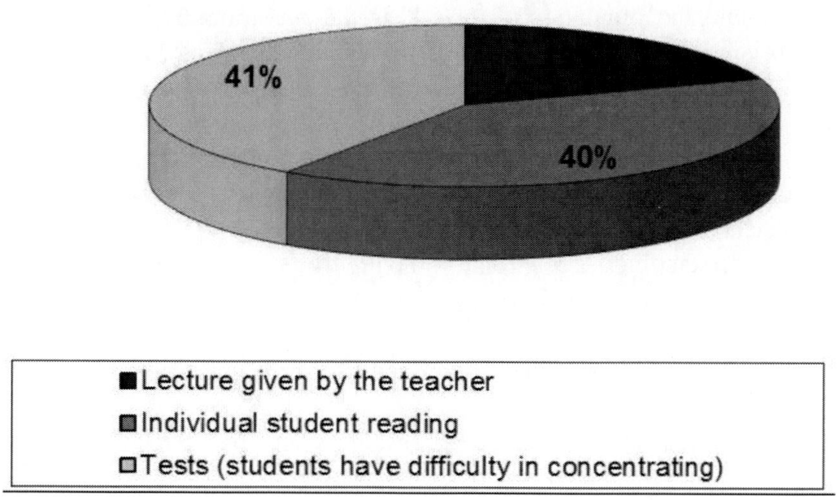

Figure 13. Responses to the question: "Which classroom activity is most affected by noise?

Similarly to the teachers, Figure 13 indicates that the students perceive that test taking, individual reading and lectures given by the teacher are activities strongly affected by noise.

The speech transmission index STI is an acoustic descriptor that considers the effects of reverberation, background noise and the contribution of the direction of the source to determine speech intelligibility. These elements, which are usually treated individually, are combined in a single index (Harris, 1998; Kang, 2002). The STI simulations were performed according to the IEC 60268-16 standard (IEC, 2003). To obtain acoustic parameters through simulations required first making a three-dimensional drawing of the room. Suitable calculation parameters were then inserted (such as the length of the impulse response), the characteristics of the finish surfaces (absorption and scattering coefficients) and the specifications of the sound source and receiver. For the source, the IEC 60268-16 standard establishes that it should be of the pointwise directional type, in order to simulate the characteristics of the human mouth. The noise generated by the source should simulate both the timbre and volume of the human voice. To obtain objective data on the speech intelligibility in classroom 01, see Figure 2, acoustic simulations of the STI were performed using Odeon 9.0 software (Odeon, 2009). The 3D models were calibrated by comparing the measured and simulated RTs. The subjective scale equivalent to the objective values of STI are in line with the IEC 60268-16:2003 standard, as indicated in Table 4.

Table 4. Quantitative and qualitative evaluation of intelligibility – IEC 60268-16 (2003)

Quantitative evaluation of intelligibility STI	Qualitative evaluation of intelligibility (Subjective impression)
0.75 – 1.00	Excellent
0.60 – 0.75	Good
0.45 – 0.60	Fair
0.30 – 0.45	Bad
0.00 – 0.30	Very bad

Figure 14 presents the simulation of the STI in the classroom 01. As can be seen in the STI map, the speech intelligibility at most of the simulated points in the classroom varies from “fair” to “bad,” according to the subjective scale of the IEC 60268-16:2003 standard (see Table 4).

The STI was not rated “good” or “excellent” at any simulated point. This negative result for the quality of intelligibility corroborates the subjective evaluation (see Figures 12 and 13) in which both teachers and students indicate that the activities – lecture classes, individual reading, correct understanding – are strongly impaired by the noise generated in the school. To this disturbance the teachers attributed the highest levels – level 4 (considerable disruption) and level 5 (intense disruption).

The low values of STI are directly related to inadequate reverberation times for classrooms, as indicated by the measurements shown in Figure 15. They also contribute negatively to the low values of STI caused by noise generated inside the classroom and in the school.

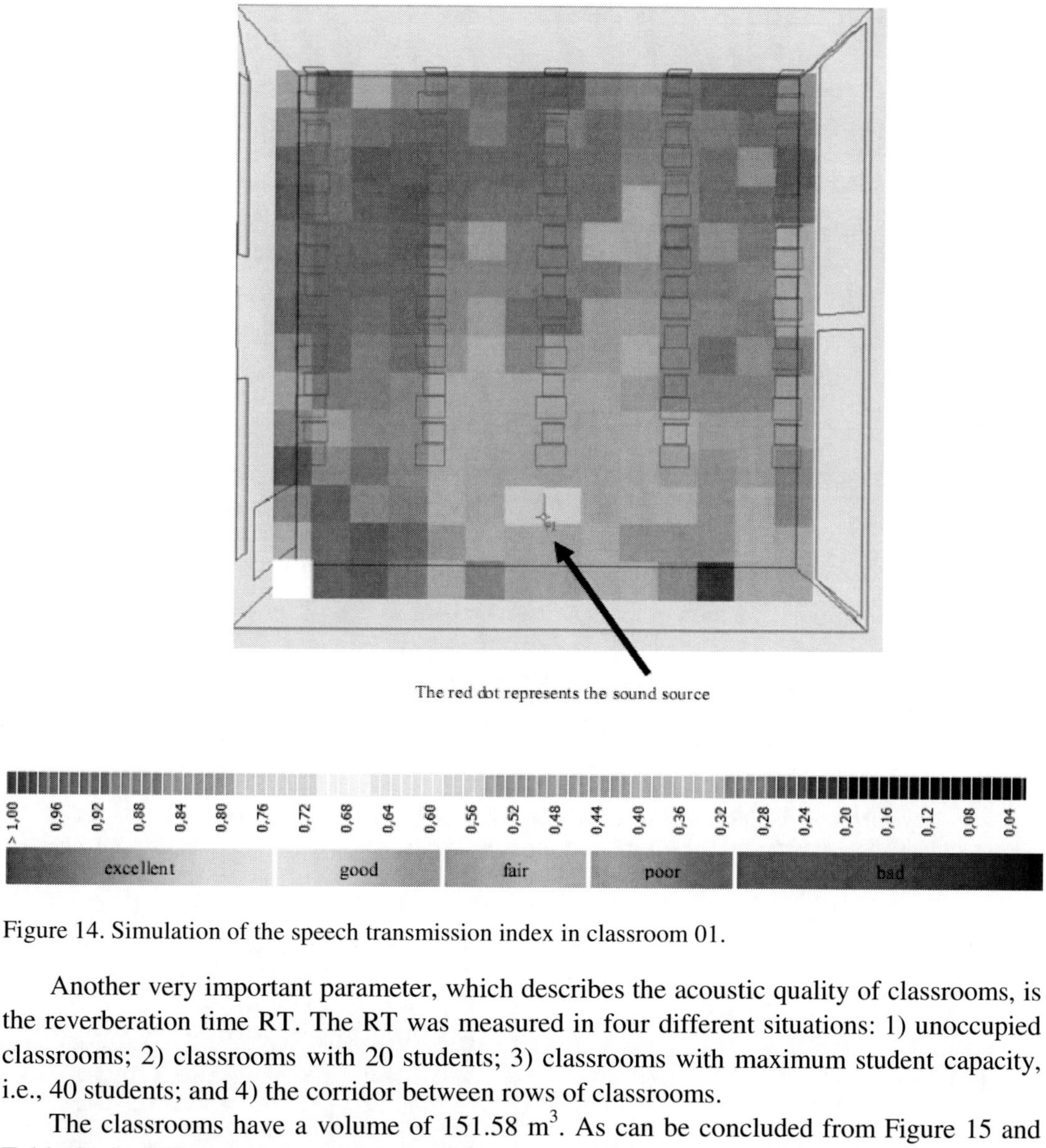

Figure 14. Simulation of the speech transmission index in classroom 01.

Another very important parameter, which describes the acoustic quality of classrooms, is the reverberation time RT. The RT was measured in four different situations: 1) unoccupied classrooms; 2) classrooms with 20 students; 3) classrooms with maximum student capacity, i.e., 40 students; and 4) the corridor between rows of classrooms.

The classrooms have a volume of 151.58 m^3. As can be concluded from Figure 15 and Table 3, the RT measurements of the classrooms in Figure 2, do not meet any of the values recommended by national and or international standards. Moreover, the measured RTs in classrooms also do not meet the recommendations of the World Health Organization, which specifies an average reverberation time of about 0.6 s.

High reverberation times are an indication of the absence of sound absorbing materials inside classrooms, resulting in poor acoustic quality. Figure 15 indicates that the classroom reaches an acceptable RT according to national and international standards only when it is occupied at full capacity, i.e., 40 students.

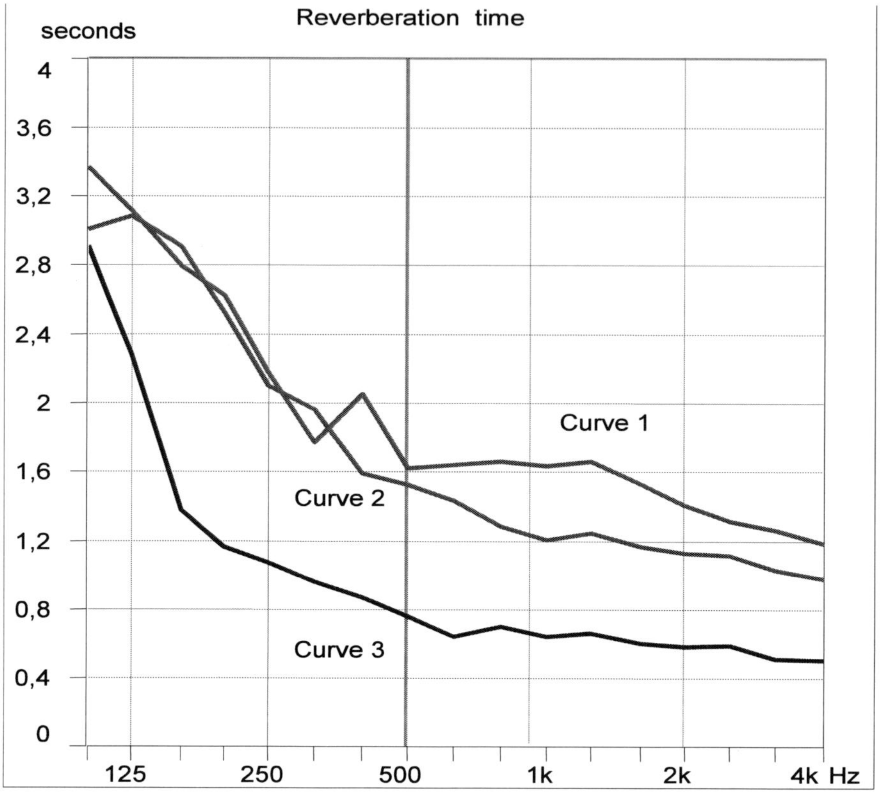

Figure 15. Reverberation time RT, as a function of occupation and frequency: Curve 1 – unoccupied classroom; Curve 2 – classroom with 20 students, or 50% of full capacity; Curve 3 – classroom with 40 students, or 100% of full capacity.

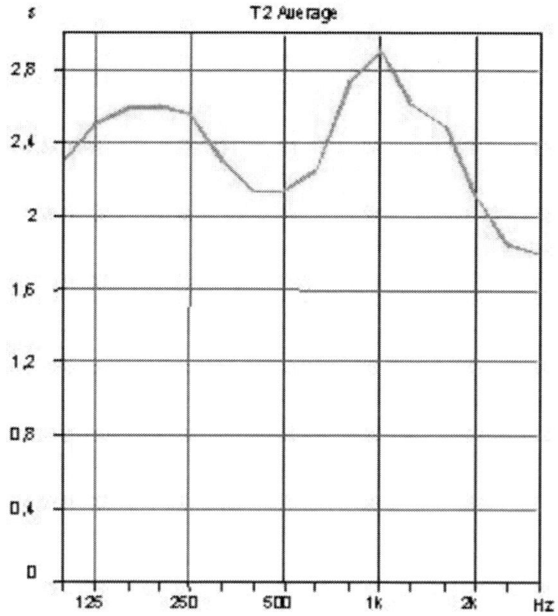

Figure 16. Measurement of RT in the corridor separating two rows of classrooms.

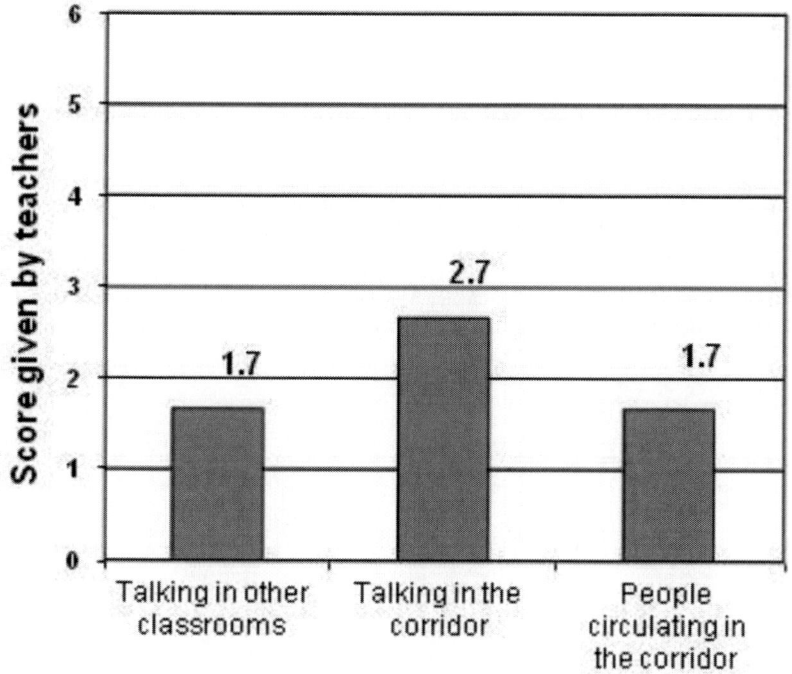

Figure 17. Responses to the question posed to the teachers: "To which factor/sound source is the noise generated outside the classroom, but inside the school building directly related?"

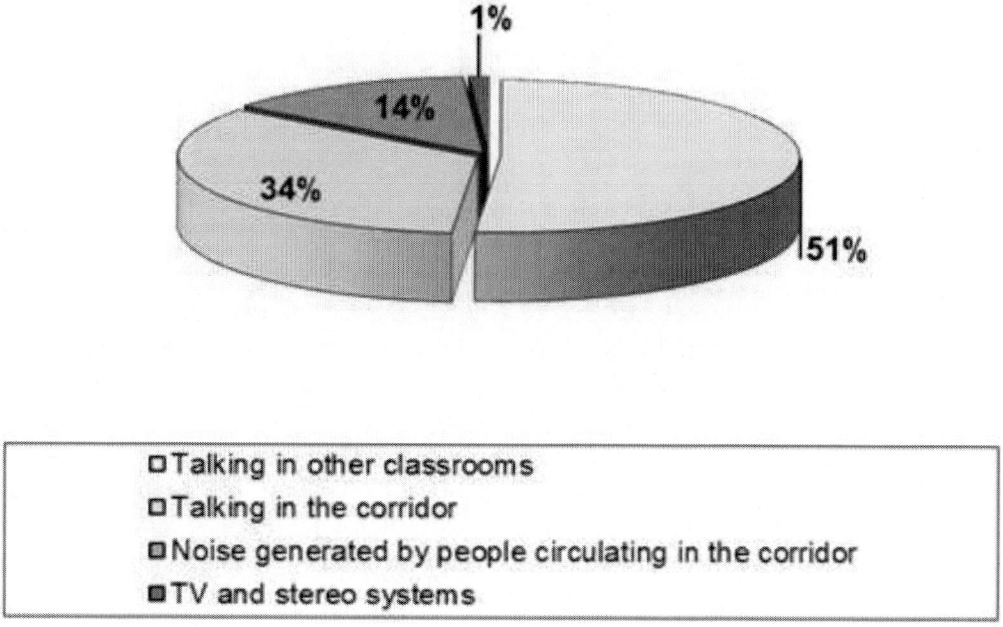

Figure 18. Responses to the question posed to the students: "To which factor/sound source is the noise generated outside the classroom, but inside the school building directly related?"

The RT was also measured in the corridor (volume of 154.4 m^3) between two rows of classrooms. The corridor was unoccupied. During this measurement, classrooms 01, 02 and

04 had classes in session while classroom 03 was unoccupied. As Figure 16 shows, the RT in the corridor was extremely high. The equivalent sound pressure level measured in the corridor without the presence of people was also very high, i.e., L_{eq} = 70.3 dB(A).

According to the teachers the noise generated outside the classroom, but inside the school building is caused by people talking in the corridor.

The students were asked the same question as that shown in Figure 18, i.e., "To which factor/sound source is the noise generated outside the classroom, but inside the school building directly related?" Figure 18 indicates that 51% of the students cited conversations in other classrooms as the main source of disruption inside the school building. In addition, 48% of the students stated that people talking and circulating in the corridor is the second greatest source of disruption inside the school building.

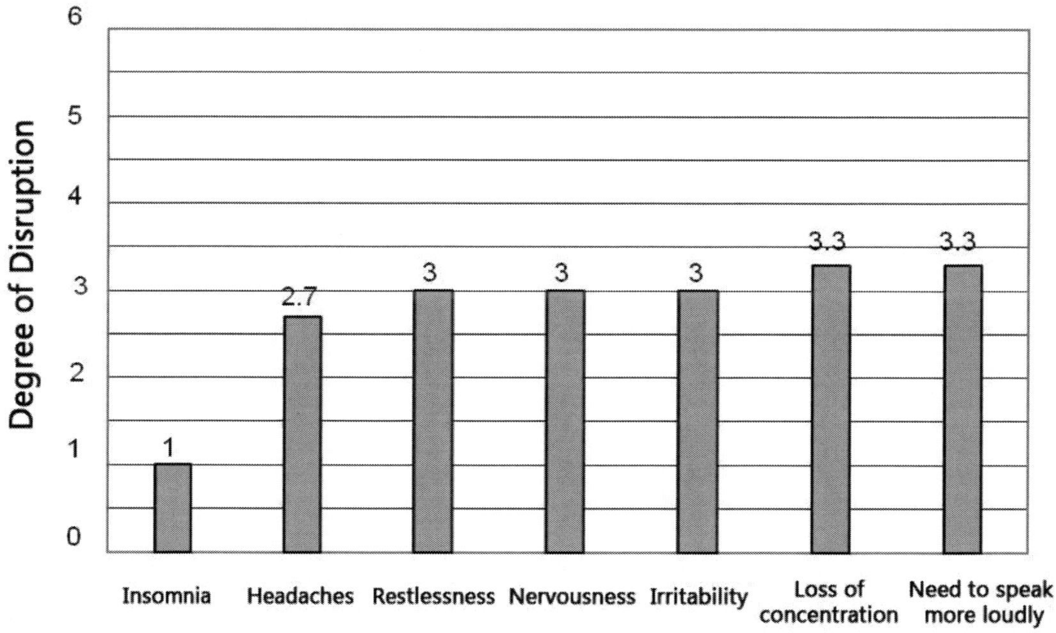

Figure 19. Teachers' responses to the question: "What effects does noise have on you?"

Figure 19 illustrates the various reactions that noise elicits in teachers. The main reactions are: 1) The need to speak more loudly, 2) Loss of concentration, 3) Irritability, restlessness and edginess, 3) Headaches, fatigue and buzzing in the ears. Over time, all these factors, separately or jointly, can have deleterious effects on the health of these individuals/workers.

The need to raise the voice is a serious occupational health that can lead to vocal fatigue. Frangos (2003) reports that, according to the National Center for Educational Statistics (USA): "*teachers miss about two days per year due to vocal fatigue, caused by raising their voices to talk over the ambient noise in and around the school classroom.*"

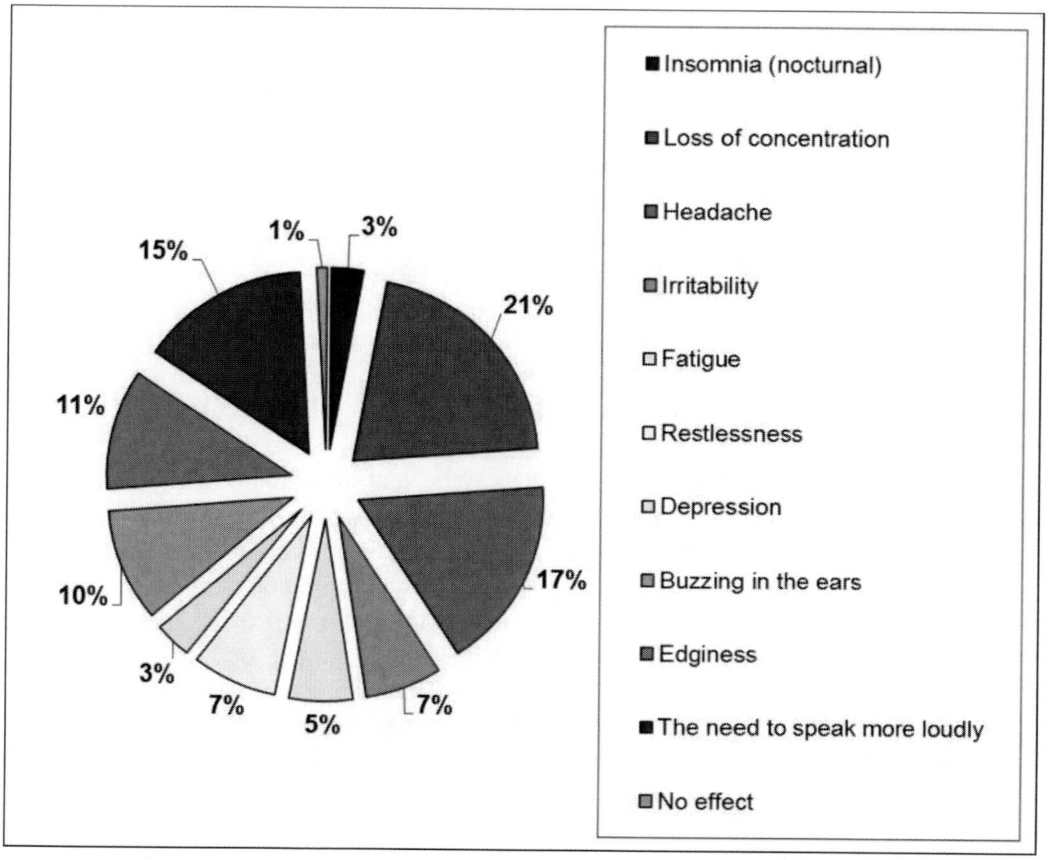

Figure 20. Students' responses to the question: "What effects does noise have on you?"

Figure 20 indicates that the four most important effects of noise on students are: 1) loss of concentration, 2) headache, 3) the need to speak more loudly, 4) edginess, 5) buzzing in the ears, 6) irritability and restlessness, and 7) fatigue.

CONCLUSION

The acoustic characteristics of classrooms were evaluated from the standpoint of ambient noise, reverberation time RT, and teachers and students' perception of noise.

The ambient noise levels measured in the classrooms exceeded those specified by the national standard, leading to the conclusion that they are not acoustically comfortable. The reason for this situation is likely the low sound insulation between the classroom and corridor, since the walls between them have permanent openings of glass blocks to provide light and ventilation. Another reason is the school's location in a neighborhood affected by external noise. This problem could be mitigated by improving the sound insulation of the school's façades.

With regard to reverberation time, the measurements revealed that the classrooms did not meet the specifications of any national or international technical standards. The RT measurements and the constructive model of the classrooms with their finishing materials clearly indicate that the classrooms, with their concrete ceilings, plastered brick walls and ceramic tile flooring, lack sound absorbing materials. The same applied to the corridor, which is finished with ceramic floor and wall tiles. There is a pressing need to reduce the reverberation time of this type of classroom, since teachers and students cited, among other negative effects of noise, the need to speak more loudly to overcome the noise inside the classroom. For teachers, in particular, this situation poses an imminent risk to their vocal health.

The assessment of teachers and students' perception of noise indicated that they are strongly affected by noise inside the school. The main reactions to noise cited by the teachers were: 1) the need to speak more loudly, 2) loss of concentration, 3) irritability, restlessness and edginess, 3) headaches, fatigue and buzzing in the ears. The students cited as their main reactions: 1) loss of concentration, 2) headaches, 3) the need to speak more loudly, 4) edginess, 5) buzzing in the ears, 6) irritability and restlessness, and 7) fatigue.

Teachers and students indicated that the main sources of noise in the school are: 1) noise generated in the classroom itself, 2) noise generated by students in other classrooms, and 3) the teacher's voice in the classroom next door. This finding points to the urgent need to improve the acoustic quality of these classrooms by reducing their reverberation time and improved sound insulation between the classrooms and corridor and sound insulation of the façades. The findings of this study lead to the conclusion that the design of buildings in general and of schools in particular should include acoustic designs, whose parameters should include reverberation time, speech transmission index and sound insulation. The noise in classrooms has extremely deleterious effects on activities such as lecture classes, individual reading, and test taking, thus impairing teacher and student performance.

ACKNOWLEDGMENTS

The authors gratefully acknowledge the German Government, through the German Academic Exchange Service – DAAD (Deutscher Akademischer Austauschdienst) and the Brazilian Government, through the National Council for Scientific and Technological Development – CNPq, for their financial support, which enabled the purchase of the sound level meters and software used in this study.

REFERENCES

ABNT – Associação Brasileira de Normas Técnicas, 1987. NBR 10152: Níveis de ruído para conforto acústico: Procedimento: [Brazilian Association of Technical Standards, 1987. NBR 10152: *Sound levels for acoustic comfort: Procedure*:]. Rio de Janeiro (in Portuguese).

Associação Brasileira de Normas Técnicas. NBR 12179: Tratamento acústico em recintos fechados: Procedimento [Brazilian Association of Technical Standards. NBR 12179: Acoustic treatment of indoor spaces: Procedure] Rio de Janeiro, 1992 (in Portuguese).

Associação Braseileira de Normas Técnicas NBR 10151: Acústica – Avaliação do ruído em áreas habitadas, visando o conforto da comunidade - Procedimento. [Brazilian

Association of Technical Standards. NBR 10151: *Noise assessment in populated areas,* aimed at *the comfort of the community*: Procedure] Rio de Janeiro, 2000 (in Portuguese).

American National Standard ANSI S12.60: Acoustical performance criteria, design requirements, and guidelines for schools. Melville, 2002.

Astolfi A, Corrado V, Fillipi M.. Classroom acoustics assessment: a procedure for field analysis. In: Euronoise Naples 2003. Italy.

Building Bulletin 93, Acoustic design of schools – a design guide. London, 2003.

DIN 1804: Acoustic quality in small to medium-sized rooms (Hörsamkeit in kleinen bis mittelgrossen Räumen). Germany, 2004.

Fundepar. Norma para escolha e aquisição de terrenos destinados a construção de prédios escolares [S*tandard for land selection and purchase for* the *construction of school buildings*] Curitiba, 1990 (in Portuguese).

Freire P. Pedagogia da indignação: cartas pedagógicas e outros escritos [*Pedagogy of indignation: pedagogical letters and other writings*] São Paulo: Editora Unesp, 2000 (in Portuguese).

Frangos A. School Districts Spend to Ensure Good Acoustics. Wall Street Journal Jul 30, pg B1. New York, N.Y. 2003.

Fukuchi T, Ueno K. Guidelines on acoustic treatments for school buildings proposed by the Architectural Institute of Japan. In: ICA – International Conference on Acoustic, Kyoto, Japan, 2004.

Harris CM. Noise Control in Buildings. McGraw-Hill, Inc. New York, USA, 1994.

Harris CM. Handbook of acoustical measurements and noise control. USA: Acoustical Society of America, 1998.

Hagen M, Hubert L, Kahlert J. Acoustic school desing. In: Forum Acusticum, 2002, Sevilha.

International Electrotechnical Commission. IEC 60268- 16: Sound system equipment- Part 16: Objective rating of speech intelligibility by speech transmission index. Switzerland, 2003.

International Organization for Standardization. ISO 3382-2: Acoustics-Measurement of room acoustic parameters - Part 2: Reverberation time in ordinary rooms. Switzerland, 2008.

International Organization for Standardization. ISO 3382-1: Acoustics — Measurement of room acoustic parameters - Part 1: Performance spaces. Switzerland, 2009.

Kang J. Numerical modeling of speech intelligibility in dining spaces. Applied Acoustics, 2002, 63: 1315-1333.

Karabiber K, Vallet M. Classroom acoustics policies – an overview." In: Euronoise Naples 2003, Italy, 2003.

Maxwell EL, Evans GW. The Effects of Noise on Preschool Children's Pre-Reading Skills. Journal of Environmental Psychology, 2000, 20: 91-97.

Martins MIM. A interferência do ruído no reconhecimento da fala: análise do ambiente e da voz do professor [*Noise interference in speech recognition*: analysis *of the environment and* of *the teacher's voice*]. In: XX Encontro da SOBRAC, II Simpósio Brasileiro de Metrologia em Acústica e Vibrações – SIBRAMA [*20th Meeting of SOBRAC, 2nd Brazilian Symposium on Acoustics and Vibration Metrology*]. Rio de Janeiro. 2002 (in Portuguese).

Lubman, D., Schuterland LC. Good classroom acoustics is a good investment. In: International Congress on Acoustics Rome, Italy, 2003.

ODEON Software Handbook - Version 9.0.

Passero CRM, Zannin PHT. Statistical comparison of reverberation times measured by the integrated impulse response and interrupted noise methods, computationally simulated with ODEON software, and calculated by Sabine, Eyring and Arau-Puchades' formulas. Applied Acoustics, 2010, 71: 1204–1210.

Santos MJ. O. Ruído no Ambiente Escolar – Causas e Conseqüências [*Noise in the School Environment - Causes and Consequences*], Universidade Federal do Rio de Janeiro [*Federal University of Rio de Janeiro*]. (RJ). 1993. 205p. (in Portuguese).

Seep B, Glosemeyer R, Hulce E, Linn M, Aytar P, Coffeen R. Classroom Acoustics: A resource for creating learning environments with desirable listening conditions. Acústica e Vibrações, 2000, 29: 2-22. (in Portuguese).

Stansfeld S. Noise exposure from various sources – cognitive effects on children. In: WHO Technical meeting on exposure response relationships of noise on health. Bonn, Germany, 2002.

World Health Organization – WHO. Noise in schools. Geneva, 2001.

Zannin PHT, Zwirtes DPZ. Evaluation of the acoustic performance of classrooms in public schools. *Applied Acoustics*, 2009, 70: 626-635.

Zannin PHT, Passero CRM, Sant'Ana DQ. Acoustic Design of Enclosed Spaces. [in] Ergonomics: Design, Integration and Implementation. Bram N. Brinkerhoff [Ed]. New York: Nova Science Publishers, 2009.

Zannin PHT, Passero CRM, Sant'Ana DQ., Bunn F, Fiedler PEK, Ferreira AMC. Classrooms Acoustics: Measurements, Simulations and Applications. [in] Classrooms: Management, Effectiveness and Challenges. Rebecca J. Newley [Ed]. New York: Nova Science Publishers, 2011.

Zannin PHT, Zwirtes DP., Passero CRM. Assessment of Acoustic Quality in Classrooms - Based on Measurements, Perception and Noise Control. [in] Noise Control, Reduction and Cancellation - Solutions in Engineering. Daniela Siano [Ed]. Croatica: InTech – Open Science. www.intechopen.com, 2012.

In: Noise and Ergonomics in the Workplace
Editor: Paulo Henrique Trombetta Zannin

ISBN: 978-1-62417-573-2

Chapter 3

ASSESSMENT OF THE QUALITY OF LIFE AND VOCAL HEALTH OF UNIVERSITY PROFESSORS

Gustavo Levandoski*[*]*, Fabricio Cieslak and Paulo Henrique Trombetta Zannin

Laboratory of Environmental and Industrial Acoustics and Acoustic Comfort,
Federal University of Paraná, Brazil

ABSTRACT

The present study examined the quality of life and vocal health of university professors. Fifty-one professors participated in the study – 25 males and 26 females. Their overall quality of life was assessed by the WHO Quality of Life-BREF instrument (WHOQOL-BREF), while their vocal health was evaluated based on the Quality of Life and Voice (QLV) questionnaire.

1. INTRODUCTION

Studies on ergonomic principles seek, among other things, to indirectly measure the load in the work environment. Among the principles of Ergonomics, the workload is considered the physical and cognitive effort required during the work activity (Wisner, 1994). Thus, Ergonomics can be considered an interdisciplinary science comprised within the areas of work physiology and psychology, which investigates the occupational environment of humans with a view to improving welfare and work performance (Grandjean and Hüting, 1977).

When possible joint factors that interfere in the work environment and impair the worker's health are analyzed, extrinsic factors are often inferred, such as a loud noise source that produces high sound levels, causing discomfort throughout the workday. However, when a professor gives classes in an environment such as the aforementioned one, it is assumed

[*] E-mail: glevandoski@gmail.com.

that, to overcome the background noise, his speech must be delivered with a higher vocal intensity, leading to fatigue vocal over time. Frangos (2003) reports that, according to the National Center for Educational Statistics (USA), teachers miss about two days per year due vocal fatigue, caused by raising their voices to talk over the ambient noise in and around the school classroom.

A new line of study that has become increasingly widespread is Voice Ergonomics, which focuses on understanding voice disorders and speech intelligibility originating from the *work environment (Vilkman, 2000; Sala et al., 2005).* A variety of meanings have been attributed to a good quality of life, about which there is a general consensus. However, different populations see quality of life as consisting of a variety of elements, such as security, safety, happiness, leisure, health, a stable financial condition, love and work, which can determine the levels of quality of life (Minayo et al., 2000; Gonçalves and Vilarta, 2004). The World Health Organization defines quality of life as "the individual's perception of his status in life, within the context of the culture and system of values in which he lives and in relation to his goals, expectations, standards and concerns" (The WHOQOL Group, 1995). The presence of the voice in the process of human socialization is one of these components, causing impacts on the quality of life of individuals, particularly on those who use their voices professionally (Penteado and Bicudo-Pereira, 2003).

The voice is a key tool in the professional life of a teacher, especially as a constitutive component of his identity, which must convince and influence his audience. His voice requires precise adaptation to avoid risking the early or eventual appearance of dysphonic symptoms (Garcia, 1986; Penteado, 1999; Penteado 2003, Penteado and Bicudo-Pereira 2003; Grillo 2004;).

The use of the voice is essential in teaching and is considered the main educational tool for the transmission of knowledge (Souza and Ferreira, 2000; Arruda, 2003; Gonçalves, 2003). As a voice professional, the teacher presents specific vocal characteristics that differentiate him from individuals who do not use their voice professionally (Behlau et al., 2005).

Through vocal quality, the professional teacher uses the resources of his vocal discourse to express his authority, thereby strongly influencing the dynamics of his relationship with his students (Dragone et al., 2006). In addition to its detrimental effect on the teacher's interaction with students, an imbalance in his vocal functioning causes changes in his vocal quality, interfering with the language performance and resulting in a negative effect on the educational process (Rogerson and Dodd, 2005).

The characteristics required for professional development in the teaching profession involve vocal demand and a moderate voice quality. These characteristics therefore reinforce the need for the teacher to have strong vocal strength to enable him to perform his daily professional activities without impairing his quality of life (Vianello, 2006). Among several physical aspects affecting the teacher's health, the voice is subjected to the structure of work organization and to certain conditioners resulting from the precariousness of the exercise of this profession (Gonçalves, 2003; Vianello, 2006).

The impact of this exercise on vocal quality is uncertain, but can range from mild to severe dysphonia. For a professional who uses his voice to perform his work, dysphonia may impair his ability to do so, since the performance of his work depends on an acceptable vocal quality (Behlau, 2006). The level of voice-related quality of life is a problem that interferes directly in the quality of life of teachers. Thus, in the following topics of this study, the quality of life and vocal health of university professors were analyzed and compared.

2. Methodological Considerations

The participants chosen for this research were 144 teachers from 12 departments of a public university located in the northwest of the state of Paraná (Brazil). A sampling loss of 64.5% occurred, resulting from teachers lacking time to participate, absence of the interviewee at the agreed upon time, and refusal to participate during the initial contact. Thus, the final sample size comprised 51 teachers, 25 males and 26 females, ranging in age from 26 to 79 years. All the participants were informed about the procedures through a Free Prior Informed Consent Form that explained the procedures and the purpose of the research, according to the "Guidelines and Regulatory Rules for Research Involving Humans" (196/96), published by the Brazilian National Health Commission and approved by the Ethics Committee of the Federal University of Paraná.

The instruments used were the WHOQOL-BREF created by the WHOQOL Group (1995), adapted and translated into Portuguese by Fleck (2000), which verifies the overall level of quality of life, plus four domains: physical, psychological, social and environmental. The second instrument was the Quality of Life and Voice (QLV) questionnaire created by Hogikyan and Sethuraman (1999) and adapted and translated into Portuguese by Behlau and Gasparini (2009), which assesses the impact of dysphonia on life, particularly that of teachers (Grillo and Penteado, 2005; Penteado and Bicudo-Pereira, 2003), dividing the analysis into the general, physical and socioemotional domains.

The WHOQOL-BREF and QLV presented an internal consistency of $\alpha=0.837$ and $\alpha=0.822$, respectively, of Cronbach's alpha coefficient (a statistical coefficient used as a test to evaluate the reliability of the scale), obtained with the data from this study. Central tendency and dispersion measures were used in the descriptive analysis of individuals and the Shapiro-Wilk test for data normality.

A non-parametric inferential analysis was then adopted, using the Mann-Whitney test in order to identify differences between the groups. The Spearman Rho correlation test was used to analyze the correlations between the variables. All the statistical processing was performed using the Statistical Package for the Social Sciences (SPSS, version 14.0) for Windows, adopting a level of significance of $p<0.05$. The following controls were used for comparison: division by gender, title, age and weekly workload (WWL) of the teachers, and vocal self-evaluation.

3. Results and Discussion

Table 1 lists the central tendency measures found in the analysis of quality of life of all the teachers. As can be seen, the overall level of quality of life using the WHOQOL-BREF was higher than the values found by Penteado and Pereira (2007) in their evaluation of high school teachers in the interior if the state of São Paulo-SP; by Fernandes and Rocha (2009) with teachers in Natal-RN, and lower values than those reported by Cieslak et al. (2011) in their evaluation of physical education students, also in the interior of the state of São Paulo. In contrast, the QLV evaluation showed higher mean values in the general and physical domains and lower values in the socioemotional domain when compared with those reported by Sevilha and Roccon (2009), who also evaluated university professors.

Table 1. General frequency of Quality of Life and respective domains

	(n= 51)	Mean	Std. Dev.	Maximum	Minimum
WHOQOL-BREF	General domain	71.7	10.9	89.4	41.3
	Physical domain	74.5	13.4	100.0	42.8
	Psychological domain	72.4	12.8	95.8	41.6
	Social relations domain	75.4	15.4	100.0	25.0
	Environmental domain	67.9	12.6	90.6	43.7
QVV	General domain	90.5	10.7	100.0	52.5
	Physical domain	97.1	6.8	100.0	68.8
	Socioemotional domain	86.1	14.4	100.0	37.5
Self Evaluation	Excellent	Very good	Good	Fair	Poor
	11 (21.6%)	11 (21.6%)	23 (45.1%)	4 (7.8%)	2 (3.9%)

*p<0.05

Table 2. Comparison of Quality of Life and Voice in relation to the sexes

		Men (n= 25)	Women (n= 26)	p
WHOQOL-BREF	General domain	74.7 ± 9.4	68.8 ± 11.5	0.050*
	Physical domain	77.0 ± 12.9	72.1 ± 13.6	0.196
	Psychological domain	75.5 ± 13.1	69.5 ± 12.1	0.100
	Social relations domain	76.3 ± 16.7	74.6 ±14.4	0.708
	Environmental domain	72.3 ± 11.3	63.7 ± 12.4	0.013*
QVV	General domain	90.7 ± 11.4	90.2 ± 10.2	0.886
	Physical domain	86.1 ± 15.4	86.0 ±13.7	0.991
	Socioemotional domain	97.6 ± 6.3	96.6 ± 7.3	0.603

*p<0.05

In the vocal self-evaluation of Sevilha and Roccon's study (2009), 9.52% of the respondents rated their voices as *Very good*, 2.85% as *Good*, 33.3% as *Fair*, and as 9:54% as *Poor*. In contrast, in the present study, some of the teachers rated their voices as *Excellent*. Most of the teachers in Pereira and Penteado's study (2007) claimed they were satisfied with their voice, and 60.9% rated their voice as *Good*, *Very good* or *Excellent*.

In Table 2, note that the "General quality of life" and the "Environmental domain" evaluated by the WHOQOL-BREF presented significant differences. Male teachers presented higher values in these two variables. Cieslak et al. (2011) evaluated physical education students and found differences between men and women in all the domains. Pereira and Cotta (2006), in their evaluation of elderly individuals in the interior of the state of Minas Gerais, found that men presented significantly higher average values in the physical, psychological and environmental domains. Cruz et al. (2011) reported the same findings in adult individuals in the city of Porto Alegre, RS, Brazil. However, in their study of teachers, Penteado and Pereira (2007) found no difference in these domains in a comparison of the sexes.

With regard to vocal health, data in the literature indicate that in self-evaluations, there is a higher occurrence of voice changes, such as tiredness when speaking, among women (Ferreira et al., 2010); but this self-reported impact is perceived similarly among men as it is among women, according to Putnoki et al. (2010). In the present study, the data point to this second finding.

In Table 3, the variables showing significant differences in the QLV were the "General quality of life and voice" and the "Physical domain." In this comparison, professors with *sensu stricto* titles presented a lower voice perception than non-titled ones.

A global analysis of the teachers' responses indicates that a large proportion of them suffer from some voice disorder. This issue deserves to be taken into account when one considers the time invested in their professional qualification.

For Pereira et al. (2003), the work-related limitations resulting from a voice disorder may impair the quality of teaching, interfere with student learning and even compromise the teacher's career and health.

In Table 4, the "General quality of life and voice" and the "Physical domain" also show significant differences. In this comparison, the younger professors of up to 39 years of age presented a lower voice perception than those older than 40.

Studies have shown that changes in the glottic configuration, especially in women, possibly due to anatomical differences associated with aging (Ortiz et al., 2004); weakening of the respiratory muscles (Soares et al., 2007) and menopause (Behlau et al., 2005), are factors responsible for changes in the vocal tract.

Table 3. Comparison of Quality of Life and Voice in relation to titles

		Non-titled (n=17)	Titled (n=33)	P
WHOQOL-BREF	General domain	71.76 ± 12.32	71.4 ± 10.2	0.912
	Physical domain	76.0 ± 13.2	73.7 ± 13.7	0.566
	Psychological domain	74.2 ± 13.7	71.2 ± 12.6	0.434
	Social relations domain	75.0 ± 15.3	75.7 ±16.0	0.873
	Environmental domain	65.6 ± 13.6	68.7 ± 11.9	0.423
QVV	General domain	94.2 ± 5.7	88.3 ± 12.3	0.027*
	Physical domain	91.4 ± 8.5	83.0 ±16.3	0.023*
	Socioemotional domain	98.5 ± 2.7	96.2 ± 8.2	0.167

*p<0.05 (Non-titled = graduate and specialist; Titled = master and doctor).

Table 4. Comparison of Quality of Life and Voice in relation to age

		Age 1 (n=19)	Age 2 (n=31)	p
WHOQOL-BREF	General domain	70.1 ± 10.9	72.3 ± 10.8	0.483
	Physical domain	74.0 ± 13.8	74.5 ± 13.4	0.904
	Psychological domain	70.1 ± 12.2	73.5 ± 13.3	0.380
	Social relations domain	73.6 ± 17.4	75.8 ± 14.0	0.638
	Environmental domain	65.4 ± 12.2	69.1 ± 12.9	0.321
QVV	General domain	85.9 ± 10.3	93.1 ± 10.2	0.020*
	Physical domain	78.9 ± 13.7	90.2 ± 13.4	0.006*
	Socioemotional domain	96.3 ± 7.5	97.5 ± 6.4	0.586

*p<0.05 (Age 1= up to 39; Age 2= above 40).

Table 5. Comparison of Quality of Life and Voice in relation to weekly workload

		40 hours (n=36)	20 hours (n=10)	p
WHOQOL-BREF	General domain	71.9 ± 11.5	69.9 ± 10.0	0.622
	Physical domain	73.9 ± 13.5	76.7 ± 13.1	0.553
	Psychological domain	73.3 ± 12.8	69.1 ± 10.0	0.345
	Social relations domain	75.9 ± 14.6	72.5 ± 18.8	0.541
	Environmental domain	67.7 ± 12.6	65.0 ± 12.6	0.553
QVV	General domain	90.9 ± 10.0	90.7 ± 7.9	0.949
	Physical domain	86.6 ± 13.4	85.8 ± 11.6	0.853
	Socioemotional domain	97.4 ± 6.8	98.1 ± 2.9	0.745

*p<0.05

Table 6. Comparison of Quality of Life and Voice in relation to voice self-evaluation

		Self-evaluation 1 (n= 22)	Self-evaluation 2 (n= 23)	p
WHOQOL-BREF	General domain	77.8 ± 6.8	67.1 ± 11.7	0.001*
	Physical domain	82.6 ± 9.5	68.7 ± 13.7	0.002*
	Psychological domain	79.7 ± 10.1	67.3 ± 13.1	0.002*
	Social relations domain	79.9 ± 10.4	73.9 ±16.9	0.229
	Environmental domain	72.3 ± 10.4	63.7 ± 13.8	0.033*
QVV	General domain	95.8 ± 4.0	89.7 ± 10.6	0.040*
	Physical domain	89.7 ± 10.6	85.3 ±13.3	0.031*
	Socioemotional domain	99.4 ± 1.8	96.4 ± 7.7	0.132

*p<0.05 (Self-evaluation 1= *Excellent* and *Very good*; Self-evaluation 2= *Good*).

Table 7. Construct comparison of the variables that measure Quality of Life and Voice

Control	Instrument	Construct	p*
SEX	*WHOQOL*	11- Do you accept your physical appearance?	0.002
		12- Do you have enough money to cover your needs?	0.008
		15- How good is your mobility?	0.016
	QVV	--------------	-
TITLE	*WHOQOL*	5- How much do you enjoy life?	0.007
		13- To what extent is the information you need on a daily basis available to you?	0.047
	QVV	1- I have difficulties in raising my voice (speaking loudly) or in being heard in noisy environments.	0.029
AGE	*WHOQOL*	--------------	-
	QVV	1- I have difficulties in raising my voice (speaking loudly) or in being heard in noisy environments.	0.047
		3- I don't know how my voice will sound when I begin to speak.	0.037
		7- I have problems at work or in developing my career (because of my voice).	0.003
WWL	*WHOQOL*	5- How much do you enjoy life?	0.012
		13- To what extent is the information you need on a daily basis available to you?	0.047
	QVV	--------------	-

*p<0.05

In contrast to the above, other studies (Boone, 1992; Ortiz et al., 2004) have found that the prevalence of complaints about tiredness when speaking is statistically significantly higher in older people. However, this finding was not confirmed in the present study. It is believed that this difference can be explained by the fact that the longer experience of most senior teachers has taught them to better manage their speaking voice.

In Penteado and Pereira's study (2007), the teachers self-evaluated their voices as *Good* or *Very good*, but their perception of the impact of their daily activities on the voice may have been misjudged, except in cases where the voice already presented a severe degree of alteration. For Camargo (2007), voice disorders are frequent in all age groups, but younger age groups show proportionally higher rates of voice disorders.

Table 5 compares the variables analyzed with respect to the teachers' weekly workload (WWL). Although the averages were higher in some domains, no significant differences were found. These findings are consistent with those of Putnoki et al. (2005). In contrast, Ferreira et al. (2008), who conducted a study of telemarketing operators, found that employees with a heavier weekly workload were significantly less satisfied with their vocal health. Souza (2000) found that teachers with a weekly workload of more than 25 hours showed more severe voice changes. Cruz et al. (2011), who examined quality of life (although the reported age distribution differed from the one described in this chapter), found statistical evidence indicating that younger individuals showed higher averages in the psychological and social domains.

Table 8. Correlation between the domains of the WHOQOL-BREF and the QVV for all the participants

	Voice Self-Evaluation	QVV General domain	QVV Physical domain	QVV Socioemotional domain
WHOQOL General domain	r =0.510 (p< 0.001)	r =0.319 (p= 0.024)	r = 0.312 (p=0.027)	n/s
WHOQOL Physical domain	r =0.522 (p< 0.001)	n/s	n/s	n/s
WHOQOL Psychological domain	r =0.508 (p< 0.001)	n/s	n/s	n/s
WHOQOL Social Relations Domain	n/s	r = 0.281 (p= 0.048)	r = 0.285 (p= 0.045)	n/s
WHOQOL Environmental domain	n/s	n/s	n/s	n/s

(r = Spearman's coefficient)

Table 6 compares the variables analyzed for voice self-evaluation. As can be seen, there is a higher proportion of teachers who self-evaluated their voice as *Excellent* or *Very good* than of those who self-evaluated their voice as *Good.*

However, the two instruments did not show a statistically significant difference between the groups in the domain of socioemotional relations. In the other domains, it can be noted that the teachers who self-evaluated their voice as *Good* present lower levels of quality of life than the teachers who self-evaluated their voice as *Excellent* or *Very good.*

Table 7 compares the construct analysis of the variables contained in the two instruments. With regard to gender, the WHOQOL-BREF instrument indicated that male teachers presented better levels of quality of life than female teachers.

It was also found that teachers who do not have a masters or doctoral degree appear to enjoy their life more, but show a greater need to receive information. The same variables were differentiated in the comparison of the weekly workload, with teachers having a higher workload showing greater enjoyment in their lives.

This can be explained by the fact that teachers with lower workloads may also have teaching jobs in other places.

With regard to the QVV, it was found that difficulties in raising the speaking voice are a problem among teachers with a masters or doctoral degree. Similarly, it was found that younger teachers present this problem more often, and also claim voice fatigue and problems of this type during the execution of their work. Table 8 shows a correlation of the domains of the WHOQOL-BREF and QLV instruments involving all the participants, using the Spearman coefficient. Albeit with low values, the general domain of the two instruments was found to be correlated. With regard to the voice self-evaluation, the results of this study corroborated those reported by Penteado and Pereira (2007). Recent research into the self-evaluation of voice perception has been highly valued, and has involved individuals with and without dysphonia with the purpose of construct validation (Behlau et al., 2009).

In studies involving teachers (Grillo and Penteado, 2005), it was found that the mean score in the global domain of the QLV was evaluated as a Good voice, demonstrating general satisfaction with voice quality. However, as can be seen in Table 6, although the teachers self-evaluated this same level of satisfaction as Good, they presented statistically lower levels of Quality of Life than the others.

CONCLUSION

The university professors evaluated in this study showed a higher mean score of quality of life and vocal health than those reported in the literature with the same sample. The statistical analysis also indicated that teachers that self-evaluated their voice as *Excellent* and *Very good* presented significantly higher levels of Quality of Life.

The comparison of sex, title, age, and weekly workload using the WHOQOL-BREF instrument showed no difference among the domains, with the exception of the variable of gender, which, in the general and environmental domains, indicated that men obtained higher averages than women. In contrast, the results of the QVV instrument indicated that only the variable of gender did not differ between the groups. It was found that, in the general and physical domains, docents with higher academic titles, i.e., Master's and Doctoral degrees, reported a lower level of voice satisfaction than the non-titled teachers and specialists. These same variables differed among the more senior teachers, who reported a better perception than the younger ones. The variable of weekly workload showed no differences, indicating the same voice perception among teachers with weekly workloads of 20 and 40 hours.

A more comprehensive analysis of the quality of life and vocal health of teachers should also include, as control parameters, acoustic descriptors such as the sound pressure level of teachers' speech, the speech transmission index STI, the reverberation time RT and the background noise of classrooms.

REFERENCES

Arruda AF. A expressividade oral das professoras: análise de recursos vocais [Oral expressiveness of teachers: analysis of vocal resources]. Master's Dissertation on Speech Therapy. Pontifical Catholic University – PUC, São Paulo, 2003. (in Portuguese).

Behlau M, Oliveira G, Santos LMA, Ricarte A. Validação no Brasil de protocolos de auto-avaliação do impacto de uma disfonia [Validation in Brazil of protocols of self-assessment of the impact of dysphonia]. *Pró-Fono R. Atual. Cient.* 2009; 21(4): 326-332. (in Portuguese).

Behlau M. Voz: o livro do especialista [Voice: the specialist's book]. vol 2. Rio de Janeiro: Revinter; 2005. (in Portuguese).

Behlau M. Rehder, M.I; Valente, O. Disfonias endócrinas [Endocrine dysphonia]. In: Behlau M, organizer. Voz: o livro do especialista. vol 2. Rio de Janeiro: Revinter; 2005. (in Portuguese).

Boone D. R. Inimigos biológicos da Voz Profissional [Biological enemies of the Professional Voice]. *Pró-Fono.* 1992; 4(2): 3-8. (in Portuguese).

Camargo MO. A percepção de professoras sobre a intensidade da própria voz [Teachers' self-perception of their voice intensity]. [dissertation] Piracicaba, Faculdade de Ciências Humanas Educação. UNIMEP, 2007. (in Portuguese).

Cieslak F, Grande AJ, Levandoski G, Gordia AP, Quadros MTB, Vilela Junior GB, Leite, N. Comparativo da qualidade de vida de acadêmicos de Educação Física de Campinas- SP e Ponta Grossa-PR [Comparison of the quality of life of Physical Education teachers in Campinas-SP and Ponta Grossa-PR]. *Revista Brasileira de Qualidade de Vida.* 2011; 3(1) 53-57. (in Portuguese).

Cruz LN, Polanczyk CA, Camey SA, Hoffmann JF, Fleck, MP. Quality of life in Brazil: normative values for the Whoqol-bref in a southern general population sample. *Qual. Life Res.* 2011, 20(7): 1123-1129.

Dragone MLS, Behlau M. A fonoaudiologia brasileira e a voz do professor - olhares científicos no decorrer do tempo [Brazilian speech therapy and the teacher's voice – scientific reviews over time]. *Fonoaudiologia Brasil.* 2006; 4(2): 6-9. (in Portuguese).

Fernandes MH, Rocha VM. Impact of the psychosocial aspects of work on the quality of life of teachers. *Rev. Bras. Psiquiatr.* 2009; 31(1):15-20.

*Ferreira LP, Akutsu CM, Luciano P, Viviano NDAG. Condições de produção vocal de teleoperadores: correlação entre questões de saúde, hábitos e sintomas vocais [*Conditions of voice production of teleoperators: correlation between health issues, habits and voice symptoms*]. Rev. Soc. Bras. Fonoaudiol. 2008, 13(4): 307-15.* (in Portuguese).

Ferreira LP, Latorre MRDO, Giannini SPP, Ghirardi ACAM, Karmann DF, Silva EE. Influence of abusive vocal habits, hydration, mastication, and sleep in the occurrence of vocal symptoms in teachers. *Journal of Voice.* 2010, 24(1): 86-92.

Fleck MPA; Louzada S, Xavier M, Chachamovich E, Vieira G, Santos L, et al. Aplicação da versão em português do instrumento abreviado de avaliação da qualidade de vida "WHOQOL-Bref" [Application of the Portuguese version of the abbreviated instrument for assessing quality of life "WHOQOL-Bref"]. *Rev. Saúde. Pública.* 2000, 34(2): 178-83. (in Portuguese).

Garcia OC, Torres RP, Shasat ADD. Disfonias Ocupacionales. Estudo de 70 casos. [Occupational Dysphonia. A study of 70 cases] *Rev. Cub. Med.;* 1986; 25:998 -1009. (in Spanish)

Gonçalves A, Vilarta R. Qualidade de vida e atividade física: explorando teoria e prática [Quality of life and physical activity: exploring theory and practice]. 1st ed. Barueri: Manole; 2004. (in Portuguese).

Gonçalves GBB. Uso profissional da voz em sala de aula e organização do trabalho docente [Professional use of the voice in the classroom and organization of teaching work]. [dissertation] Belo Horizonte, Faculdade de Educação. UFMG, 2003. (in Portuguese).

Grillo MHMM, Penteado RZ. Impacto da voz na qualidade de vida de professores(as) do ensino fundamental [Impact of the voice on the quality of life of elementary school teachers]. *Pró-fono R. Atual. Cient.* 2005, 17(3):321-330. (in Portuguese).

Grillo MHMM. The impact of a vocal improvement course in a speech language and hearing science prevention context. *Pró-Fono Revista de Atualização Científica.* 2004, 16(2): 159-168. (in Portuguese).

Grandjean E, Hünting W. *Fitting the task to the man: a textbook of occupational ergonomics. Applied Ergonomics, 1977, 8(3): 135–140.*

Hogikyan, N. D.; Sethuraman, G. Validation of an instrument to measure Voice-Related Quality of Life (VRQOL). *J. Voice*, 1999, 13(4): 557-569.

Minayo MCS, Hartz ZMA, Buss PM. Qualidade de vida e saúde: um debate necessário [Quality of life and health: a necessary debate]. *Ciênc. Saúde Coletiva,* 2000, 5(1): 7-18. (in Portuguese).

Ortiz E, Lima EA, Costa EA. Saúde Vocal de Professores da Rede Municipal de Ensino da Cidade de São Paulo [Vocal Health of Teachers of the Municipal School System of São Paulo]. *Rev. Bras. Med. Trab.* 2004, 2(4): 263-6. (in Portuguese).

Penteado RZ, Bicudo-Pereira IMT. Avaliação do impacto da voz na qualidade de vida de professores [Assessment of the impact of voice on the quality of life of teachers]. *Rev. Soc. Bras. de Fonoaudiologia*, 2003, 8(2):19-28. (in Portuguese).

Penteado RZ, Pereira IMTB. Qualidade de vida e saúde vocal de professores [Quality of life and vocal health of teachers]. *Rev. Saúde. Pública.* 2007, 41(2): 236-43. (in Portuguese).

Penteado RZ. Aspectos de qualidade de vida e de subjetividade na promoção da saúde vocal do professor [Aspects of quality of life and subjectivity in the promotion of the teacher's vocal health]. [Thesis]. São Paulo. Faculdade de Saúde Pública da Universidade de São Paulo, 2003. (in Portuguese).

Pereira IMTB, Penteado RZ,Bydlowski CR, Elmor MRD, Guazelli ME. Escolas Promotoras de saúde: onde está o trabalhador professor? [Health Promoting Schools: Where is the working teacher?] *Saúde em Revista UNIMEP.* 2003, 5(11): 29-34. (in Portuguese).

Pereira RJ, Cotta RMM, Franceschini SCC, Ribeiro RCL, Sampaio RF, Priore SE, Cecon PR. Contribuição dos domínios físico, social, psicológico e ambiental para a qualidade de vida global de idosos [Contribution of the physical, social, psychological and environmental domains to the overall quality of life of the elderly]. *Rev. Psiquiat.,* 2006, 28(1): 27-38. (in Portuguese).

Putnoki DS, Hara F, Oliveira G; Behlau M. Qualidade de vida em voz: o impacto de uma disfonia de acordo com gênero, idade e uso vocal profissional [Quality of life of the voice: the impact of dysphonia according to gender, age and professional use of the voice]. *Rev. Soc. Bras. Fonoaudiol.* 2010, 15(4): 485-90. (in Portuguese).

Rogerson J, Dodd B. Is there an effect of dysphonic teachers on children's processing of spoken language? *Journal of Voice.* 2005, 19(1): 47-60.

Servilha EAM, Roccon PF. Relação entre voz e qualidade de vida em professores universitários [Relationship between voice and quality of life of university teachers]. *Rev. CEFAC.* 2009, 11(3): 440-448. (in Portuguese).

Sala E, Sihvo M, Laine A. Röstergonomi – Rösten ett fungerande arbetsredskap. Institutet för arbetshygien, Arbetarskyddscentralen, Helsingfors, Finland. *2005.*

Soares EB, Borba DT, Barbosa TK, Medved DM, Montenegro ACA. Hábitos vocais em dois grupos de idosos [Vocal habits in two groups of elderly]. *Rev. CEFAC.* 2007, 9(2): 221-7. (in Portuguese).

Souza TMT, Ferreira LP. Caracterização vocal dos professores dos municípios de São Paulo – DREM 5 [Voice characterization of teachers in the municipalities of Sao Paulo –

DREM 5]. In Ferreira, LP and Costa, HO. Voz ativa – falando sobre o profissional da voz [Active voice – speaking about the voice professional]. São Paulo: *Roca*, 2000, p. 145-162. (in Portuguese).

The Whoqol Group. The World Health Organization Quality of Life Assessment (WHOQOL): position paper from the World Health Organization. *Soc. Sci. Med.* 1995, 41(10): 1403-9.

Vianello L. O uso da voz em sala de aula: o caso dos professores readaptados por disfonia [Use of the voice in the classroom: the case of teachers readapted through dysphonia] 2005-2006. [dissertation] Belo Horizonte, Faculdade de Saúde Pública. *UFMG*, 2006. (in Portuguese).

Vilkman, E. Os problemas de voz no trabalho: um desafio para a segurança e saúde [Voice problems at work: a challenge to safety and health]. *Arranjo Folia Phoniatr. Logop.*, 2000, 52 (1-3): 120-125. (in Portuguese).

Wisner, A. A inteligência no trabalho: textos selecionados de ergonomia [Intelligence at work: selected texts on ergonomics]. São Paulo: *Fundacentro,* 1994. (in Portuguese).

In: Noise and Ergonomics in the Workplace
Editor: Paulo Henrique Trombetta Zannin

ISBN: 978-1-62417-573-2

Chapter 4

ASSESSMENT AND CORRECTION OF ACOUSTIC DESCRIPTORS IN CLASSROOMS

Paulo Henrique Trombetta Zannin**, *Elaine Carvalho da Paz* and *David Sant'Ana de Queiroz
Laboratory of Environmental and Industrial Acoustics
and Acoustic Comfort, Federal University of Paraná, Brazil

ABSTRACT

This chapter describes real examples in which several procedures were used to assess and/or correct the reverberation time and the speech transmission index in small classrooms. These procedures included in situ measurements, statistical analysis and computer simulation. The objects of study were four models of university classrooms that differ in terms of their volume, dimensions, occupation capacity and wall, floor and ceiling finishing materials. A statistical comparison was made of the measured values against reverberation times calculated by the Sabine, Eyring and Millington-Sette equations. The statistical results indicate that the reverberation times calculated by Sabine's equation were closest to the measured values considering the classroom models of this study. Computer simulations indicated that the reverberation time and speech transmission index in existing buildings can be corrected easily by acoustically treating the ceilings and floors of the classrooms.

1. INTRODUCTION

The acoustic characteristics of closed environments can be perceived daily in many different situations, particularly when verbal communication is the characteristic that defines the functionality of the environment, for example, a classroom (Hodgson, 2003).

* Email: paulo.zannin@pesquisador.cnpq.br, paulo.zannin@gmail.com.

Research has shown that acoustics affect the learning ability and performance of pupils, as well as their social behavior and interaction. Unfavorable acoustic conditions make learning and teaching exhausting (Lubman and Sutherland, 2003). According to Hodgson (2003), adequate reverberation time contributes to better learning; however, this characteristic has frequently been neglected. The study of reverberation times in classrooms is a very important theme that has been a frequent subject of numerous studies (Astolfi et al., 2008; Zannin and Zwirtes, 2009, Passero and Zannin, 2010).

Reverberation time (RT) values can be obtained through mathematical equations and measurements with adequate instrumentation (Passero and Zannin, 2010). The best known equation for calculating RT was developed by the great pioneer of modern room acoustics, physicist Wallace Clement Sabine, who introduced the notion of "Reverberation Time" around the year 1900. Sabine's equation takes into account the volume of the room, the surface area of the constructive elements and their respective absorption coefficients (Fasold and Veres, 1998):

$$RT = \frac{0.163.V}{A + 4mV} \text{ [s]} \qquad (1)$$

where:

V is the volume of the room in [m^3],

A is the equivalent sound absorption area $A = \sum_{i=1}^{n} \alpha_i \cdot S_i$, where is the area of each material, and is the absorption coefficient of the materials of the room's surfaces, $4mV$ corresponds to the sound absorption by air, expressed in [m^2], and is the energy attenuation coefficient of air, expressed in [10^{-3} m^{-1}].

According to Eyring (1930), the sound absorption coefficient proposed by Sabine, which considers a uniform distribution of energy, would not be a coefficient of real absorption but a logarithmic function representing the variations of the surface absorption of all the elements in the room. Hence, a new formula for predicting the RT, called the "Eyring Formula," was developed and is presented below:

$$RT = \frac{0.163.V}{-S.\ln(1-\overline{\alpha})} \text{ [s]} \qquad (2)$$

where:

RT is the reverberation time in [s],

V is the room's volume in [m^3],

S is the total area of the room's surfaces in [m^2],

$\ln$ is the Neperian logarithm, and

$\overline{\alpha}$ is the mean absorption coefficient of all the materials: $\overline{\alpha} = \frac{1}{S}\sum_i S_i.\alpha_i$, where S_i is the area of each surface that makes up the room, and α_i is the sound absorption coefficient of those materials.

Similarly to the concept developed by Eyring, based on Sabine, Millington (1932) and Sette (1933), a third formula was developed that differs from Eyring's concept in that it considers the sound absorption coefficients separately from the room's constituent elements. This third formula is shown below:

$$RT = \frac{0.163.V}{-\sum_{i=1}^{n}[S_i.\ln(1-\alpha_i)]} \text{ [s]} \tag{3}$$

where:

RT is the reverberation time in [s],

V is the room's volume in [m^3],

S_i is the area of each surface that makes up the room, in [m^2],

ln is the Neperian logarithm, and

α_i is the absorption coefficient of the materials of the room's surfaces.

Studies developed for rooms using the aforementioned formulas have demonstrated their applicability in different situations. Sabine's formula has been applied to rooms called "*live rooms*" (Eyring, 1930), i.e., closed environments that need active and mobile acoustics, such as verbal communication. The opposite of these environments are "*dead rooms,*" where Eyring and Millington-Sette's formulas are more applicable (Eyring, 1930; Millington, 1932; Sette, 1933; Neubauer, 2000).

Thus, this work uses these classic equations to predict the RT in university classrooms and to compare the results with measured reverberation times. Computer simulations were also performed in this work to correct the reverberation time and speech transmission index in a real case.

2. Assessment of Acoustic Quality – Measurements, Calculations, Statistical Analysis and Simulations

Fitzroy (1959) recommends that different types of room configurations be considered when predicting RT in rectangular rooms with volumes exceeding 250 m^3. Therefore, three different models of classrooms found in the Federal University of Paraná were selected for this study. The classrooms under evaluation were furnished but unoccupied, i.e., without the presence of students. A fourth classroom model was also evaluated by means of computer simulations to correct the reverberation time and speech transmission index.

The model R01 classrooms, Figure 1, have wood flooring, painted concrete walls, ceilings finished in wood wool, some of them partially dropped, glass windows and wooden doors, a calculated volume of approximately 367.20 m^3, and a seating capacity of 65.

The model R02 classrooms, Figure 2, which are mini auditoriums, have wood flooring, painted concrete walls, ceilings finished in wood wool, glass windows and wooden doors, a calculated volume of 294.74 m^3, and a seating capacity of 60.

The model R03 classroom, Figure 3, has wooden flooring, painted concrete walls, ceilings finished in wood wool, part of them dropped horizontally and part inclined, glass windows and wooden doors, a calculated volume of 372.30 m^3, and a seating capacity of 60.

According to the ISO 3382-1:2009 (ISO, 2009) and ISO 3382-2:2008 (ISO, 2008) standards, RT can be measured by the interrupted noise method and by the integrated impulse response method.

Figure 1. Model R01 Classroom.

Figure 2. Model R02 Classroom.

Figure 3. Model R03 Classroom.

Measuring RT by the interrupted noise method, as described in this paper, consists of exciting the room with a pseudo-random pink noise and calculating the RT from the room's response to this excitation. A common setup to measure the RT by this method comprises: 1) an omnidirectional sound source, 2) a sound power amplifier, 3) a noise generator, 4) omnidirectional microphones, and 5) a sound decay recorder and analyzer. RT was measured with a dual-channel Brüel and Kjaer BK 2260 real-time sound analyzer, a BK 2716 power amplifier, and a BK 4296 dodecahedron loudspeaker (Zannin et al., 2008; Zannin et al., 2012). The sound thus generated was captured by a microphone connected to the BK 2260 analyzer, which calculated the reverberation time for each frequency of the spectrum of interest.

Figure 4. Model R04 classroom with volume of 277.49 m^3

The sound absorption coefficients of the constituent elements of the classrooms evaluated here were taken from tables reported in the literature, which are considered internationally recognized values (Harris, 1998; Jenisch et al., 2002; Hohmann, 2004; Long, 2006). The RTs of each classroom were then calculated using Sabine, Eyring and Millington-Sette's formulas.

The results of these calculations were subjected to a comparative analysis with the measured values in order to identify the most applicable formula for predicting RT in the classroom models in question. This comparative analysis was performed in two ways: (1) graphically, based on reverberation time curves as a function of frequencies; and (2) statistically, based on an analysis of the linear relationship between predicted and measured values. It should be noted that none of the calculations considered the presence of people in the classrooms, since one of the parameters of comparison was the values of the measurements taken in unoccupied classrooms.

To analyze and correct the reverberation time and speech transmission index, another classroom model was evaluated – model R04, which is depicted in Figure 4.

The R04 model has the following characteristics: classrooms designed in modules, with reinforced concrete structure (columns, beams and slabs); plastered and painted masonry walls; PVC window frames and ordinary 4 mm thick glass panes; painted wooden doors; unplastered ribbed roof slab with exposed beams, and vinyl flooring tiles in the classrooms and passageways. Another major difference between the 4 classrooms models is that the R01, R02 and R03 models were built in the 1960s, while R04 classroom model started to be built in the year 2000.

The classroom evaluated here has a volume of $V = 277.49\ m^3$ and a seating capacity of 40. The computer simulations to correct the reverberation time and speech transmission index were performed using Odeon v.9.0 software (ODEON, 2007).

3. Results and Discussion

Figure 5 illustrates the RT curves as a function of frequency for the predicted (calculated) and measured RTs for model R01 classrooms. An analysis of this figure reveals a tendency among the curves of the RT values predicted by the Sabine and Millington-Sette formulas, as well as a relative correlation of these curves with those of the measured values.

Figure 6 shows the RT curves as a function of frequency of the predicted and measured values for model R02 classrooms. An analysis of Figure 6 reveals a similar trend for the model R01 classrooms; however, the best adjustment of the curves at low frequencies of 125 Hz to 500 Hz show a correlation between the curve of the measured RT values and the curves of the RT values predicted by the Sabine and Millington-Sette formulas.

Figure 7 shows the RT curves as a function of frequency for predicted and measured values for model R03 classrooms. An analysis of Figure 7 indicates a tendency among the curves of the RT predicted by Sabine's formula and the curve of the RT values measured at low frequencies of 125 Hz to 500 Hz.

Figure 8 and Table 1, respectively, present dispersion graphs and calculated coefficients of correlation of samples of predicted and measured values of RT.

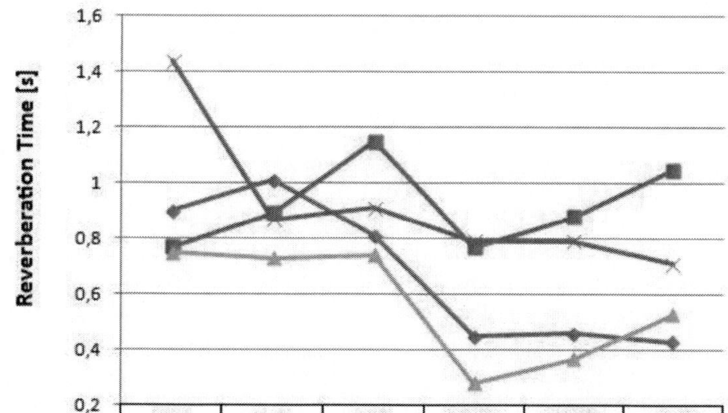

	125	250	500	1000	2000	4000
Sabine	0,9	1,01	0,81	0,45	0,46	0,43
Eyring	0,77	0,89	1,15	0,77	0,88	1,05
Millington-Sette	0,75	0,73	0,74	0,28	0,37	0,53
Measured	1,44	0,87	0,91	0,79	0,79	0,71

Frequency [Hz]

Figure 5. Calculated and measured reverberation times for model R01 classrooms.

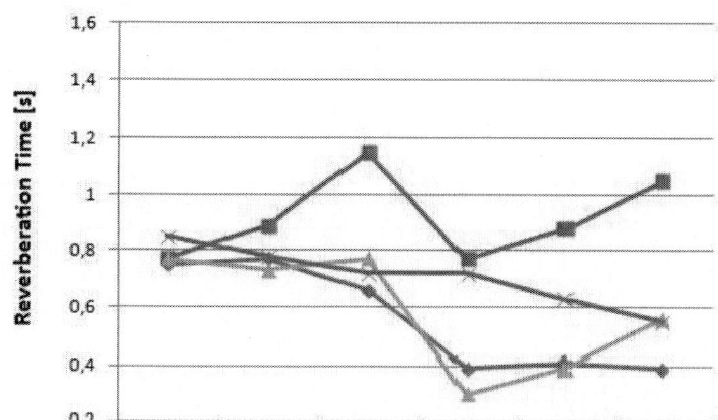

	125	250	500	1000	2000	4000
Sabine	0,75	0,77	0,66	0,39	0,41	0,39
Eyring	0,77	0,89	1,15	0,77	0,88	1,05
Millington-Sette	0,77	0,73	0,77	0,3	0,39	0,56
Measured	0,85	0,78	0,72	0,72	0,63	0,55

Frequency [Hz]

Figure 6. Calculated and measured reverberation times for model R02 classrooms.

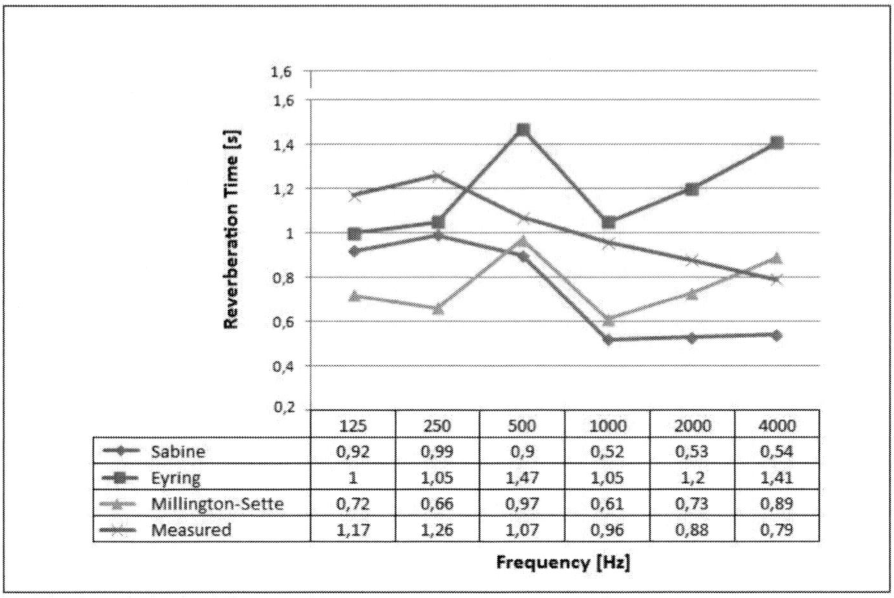

	125	250	500	1000	2000	4000
Sabine	0,92	0,99	0,9	0,52	0,53	0,54
Eyring	1	1,05	1,47	1,05	1,2	1,41
Millington-Sette	0,72	0,66	0,97	0,61	0,73	0,89
Measured	1,17	1,26	1,07	0,96	0,88	0,79

Figure 7. Calculated and measured reverberation times for model R03 classrooms.

An analysis of Figure 8 and Table 1 indicate that the sample of values predicted by Sabine's formula presents the best correlation with the sample of measured values.

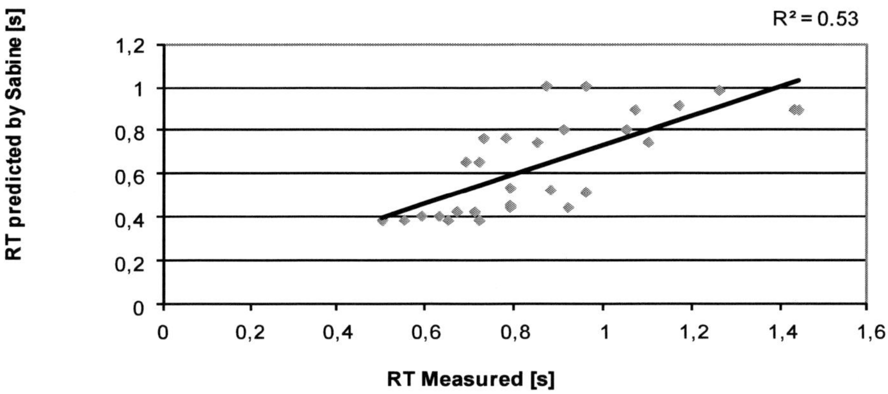

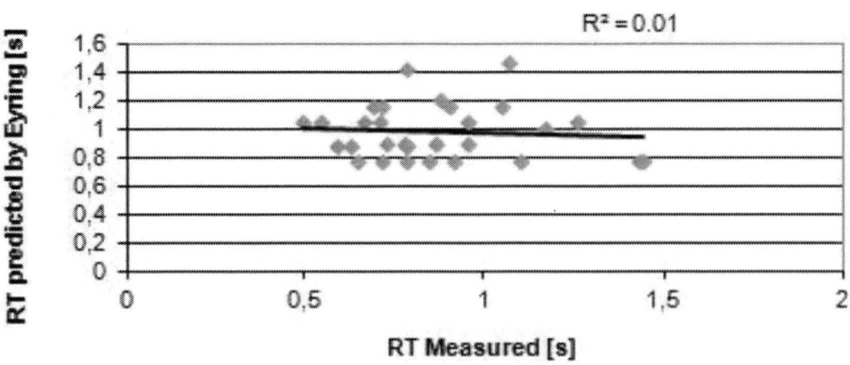

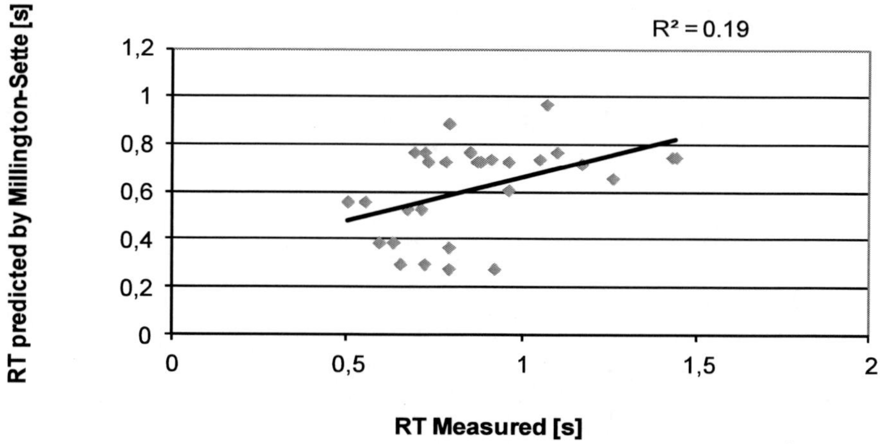

Figure 8. Dispersion graphs for predicted (calculated) and measured Reverberation Time RT.

Table 1. Linear correlation coefficients

Relation	Correlation Coefficient	Classification
Sabine vs. Measured	0.73	Strong and Direct
Eyring vs. Measured	-.08	Weak and Inverse
Millington-Sette vs. Measured	0.44	Average and Direct

Table 2 lists the values of reverberation time, RT, measured in the model R04 classroom.

Table 2. Reverberation time RT of the model R04 classroom

Classroom model	125 Hz	250 Hz	500 Hz	1000 Hz	2000 Hz	4000 Hz
RT - R04	3.6 s	3.7 s	2.7 s	2.6 s	2.2 s	1.7 s

The Brazilian NBR 12179 (1992) standard for the acoustic conditioning of rooms recommends RT ~ 0.6 s for the frequency of 500 Hz and V = 277.49 m^3. It should be kept in mind that the RT values established by the Brazilian standard are actually for conference rooms, but this type of room is the only one in the standard that resembles that of classrooms. Taking the French standard cited in WH0 (2001), the RT (obtained from an arithmetic average for the frequencies of 500, 1000 and 2000 Hz) recommended for rooms with V > 250 m^3, should fall within an interval of 0.6 < RT 1.2 s.

Considering these examples of national technical standards and examining Table 2, it is clear that the R04 model presents an extremely poor reverberation time for classrooms. At the frequency of 500 Hz, the RT of the room is 2.7 s, in other words, much higher than the RT ~ 0.6 recommended by the Brazilian standard. According to the French standard, the mean value of RT for the frequencies of 500, 1000 and 2000 Hz would be RT_{mean} = 2.5 s, which is also much higher than the recommended value of 1.2 s. Inadequate reverberation times can be considered a critical factor contributing to the decline in the quality of education, since communication becomes exceeding difficult, compromising the teaching and learning process.

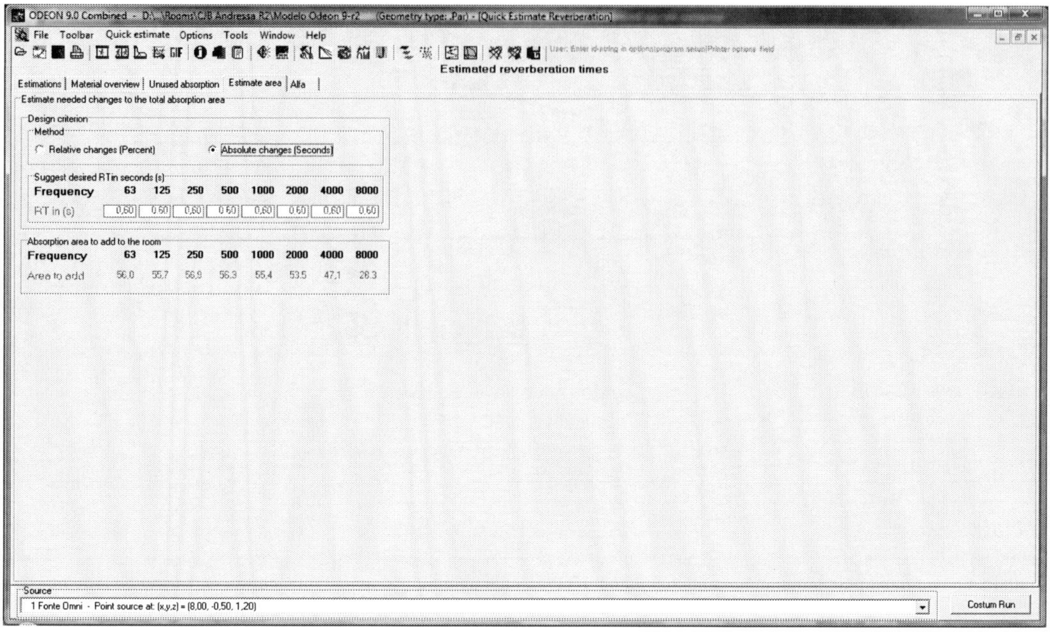

Figure 9. Calculation of absorption areas as a function of desired reverberation time.

Table 3. Sound absorption coefficients of the ceiling and floor in classroom R04

Sound absorption coefficient α	125 Hz	250 Hz	500 Hz	1 kHz	2 kHz	4 kHz
Ceiling – ribbed slab	0.01	0.01	0.015	0.015	0.02	0.02
Flooring – vinyl tiles	0.02	0.03	0.04	0.05	0.05	0.10

The simulations with Odeon software and Sabine's equation, Figure 9, indicate the need to add approximately 56.3 sabines at the frequency of 500 Hz, 55.4 sabines at the frequency of 1000 Hz, 53.5 and sabines at the frequency of 2000 Hz in order to reach the reverberation time of 0.6 s in the model R04 classroom.

Table 3 presents the sound absorption coefficients of the ceiling and floor in classroom R04 (Harris, 1998; Jenisch et al., 2002; Fasold and Veres, 2003; Hohmann, 2004; Long, 2006).

The use of light porous materials, which are good acoustic absorbents to reduce reverberation and control noise in work environments, like the school environment, is a widely recommended practice among researchers and the civil construction industry.

Absorbent materials, particularly on ceilings and floors, should be used for the acoustic treatment of classrooms (Hodgson, 2002; Frangos, 2003).

To correct the RT from 2.5 s to 0.6 s, materials currently available in the civil construction market were used, as indicated in Table 4.

Table 4. Sound absorption coefficients of materials for the correction of RT

Sound absorption coefficient α	125 Hz	250 Hz	500 Hz	1 kHz	2 kHz	4 kHz
Prisma Decor ceiling – Isover do Brasil[1]	0.64	0.72	0.66	0.84	0.91	0.90
Wood parquet flooring over concrete[2]	0.20	0.15	0.10	0.10	0.05	0.10

Source: 1) ISOVER do Brasil; 2) ODEON A/S.

Table 5. Quantitative and qualitative assessment of intelligibility – IEC 60268-16 (2003)

Quantitative assessment of intelligibility STI	Qualitative assessment of intelligibility (Subjective impression)
0.75 – 1.00	Excellent
0.60 – 0.75	Good
0.45 – 0.60	Satisfactory
0.30 – 0.45	Poor
0.00 – 0.30	Extremely poor

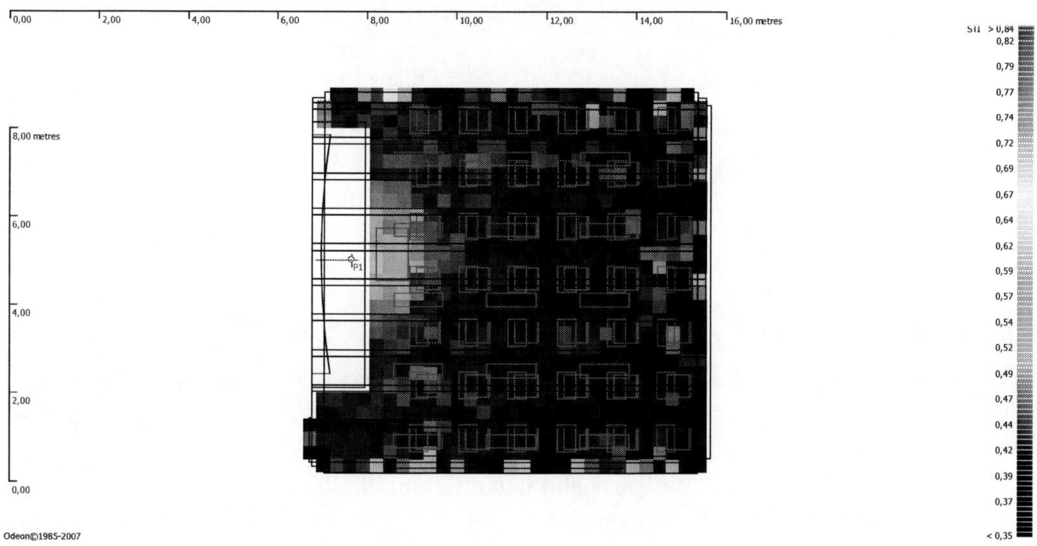

Figure 10. STI map for classroom R04.

Another very important descriptor of the acoustic quality of a classroom – the speech transmission index, STI – was studied by computer simulation for the R04 model. The speech transmission index, STI, is an acoustic descriptor that considers the effects of reverberation, background noise and the contribution of the direction of the source to determine speech intelligibility. These elements, which are usually treated individually, are combined in a single index (Harris, 1998; Kang, 2002). The STI simulations were performed according to the IEC 60268-16 standard (IEC, 2003).

The IEC 60268-16 standard (IEC, 2003) provides not only a quantitative evaluation of the STI but also a corresponding qualitative assessment, as indicated in Table 5.

Table 6. Nominal STI values

Acoustic descriptor	Point 1	Point 2	Point 3	Point 4	Point 5
STI	0.4	0.42	0.39	0.39	0.39

Table 7. Nominal STI values after replacing floor and ceiling materials

Acoustic descriptor	Point 1	Point 2	Point 3	Point 4	Point 5
STI	0.76	0.75	0.73	0.75	0.73

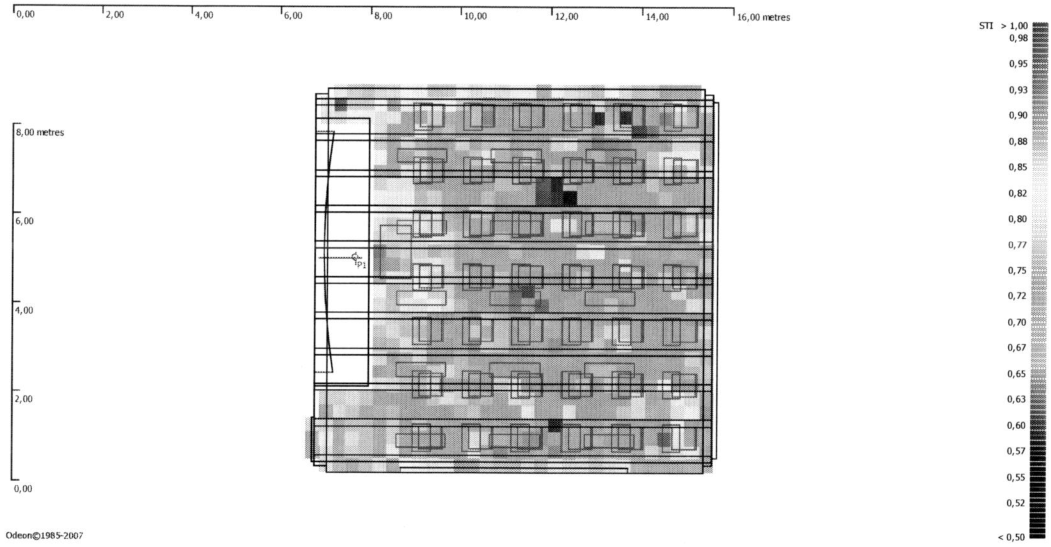

Figure 11. STI map for classroom R04 after replacing the floor and ceiling materials.

The STI was simulated for the R04 classroom model using Odeon software, as illustrated in Figure 10. The nominal STI values for five points distributed uniformly inside the room were extracted from this simulation.

Figure 10 and Tables 5 and 6 clearly show that the intelligibility in the classroom can be classified as *Poor*. The STI was also simulated after replacing the ceiling and floor finishing materials, as described earlier herein for the correction of the RT. Figure 11 and Tables 5 and 7 clearly indicate that the quality of the intelligibility, previously classified as *Poor*, changed to *Good*.

CONCLUSION

This chapter demonstrated that the evaluation and acoustic treatment of enclosed spaces in general and classrooms in particular can be performed successfully using different analytical techniques involving measurements, calculations of RT using classical equations, and computer simulations.

Four models of classrooms found in a Brazilian university were studied here. *In situ* measurements of the RT were taken according to the international standards ISO 3382-1 and ISO 3382-2. The RT was also calculated by Sabine, Eyring and Millington-Sette's equations. The RT and STI of a model of a real classroom were corrected with the help of simulations using ODEON software.

The linear correlations indicated that Sabine's formula presented a 73% correlation between the measured and calculated values, which is considered high, thus demonstrating that this formula is the most suitable one for calculating the RT of these classroom models.

Based on the above results, it is evident that reverberation time measurements produce reliable results, provided they are taken according to the recommendations of standards such as the international ISO 3382-1 and 3382-2. To obtain reliable results from these measurements, it is also necessary that the evaluators be experienced and that the equipment employed be of high quality, such as for example the devices described in the Materials and Methods section. If such measurements cannot be taken due to lack of measuring equipment, Sabine's equation can be used for classrooms with designs similar to those of this study, since this equation showed the best correlation between measured and calculated values of RT, i.e., 73%. In general, the classical equations comprise a reliable and low-cost frame of reference for predicting the reverberation time in classrooms. Another way described here to analyze the acoustic quality of classrooms is by computer simulation. The real example presented here is a room whose RT was 2.5 s, which is excessively high for the functions of a classroom. The simulations demonstrated that simply by replacing the ceiling and flooring with materials with better acoustic performance reduced the RT from 2.5 s to 0.6, i.e., a RT value compatible with a classroom, according to international standards of acoustic quality.

Also using computer simulations, it was found that the classroom with RT = 2.5 s presented STI of 0.30 to 0.45, which is *Poor* according to the IEC classification, Table 5. The substitution of the ceiling and flooring for more sound absorbent materials revealed a new STI value, i.e., STI ~ 0.75, which corresponds to the IEC classification for *Good* intelligibility. Therefore, it can be concluded that classrooms with extremely poor RT and STI are only possible when construction specifications are not accompanied by an acoustic design and/or when there is a lack of resources for the implementation of acoustic treatment of enclosed environments. Otherwise, from the standpoint of technology and know-how, every tool is available to designers. Educational environments of the most varied types – from kindergarten to universities – with deficient acoustics will inevitably generate serious social costs in the teaching and learning process. Therefore, it is to be hoped that public officials never cease to focus their efforts on allocating resources to ensure the minimum requirements for good acoustics in the classroom.

ACKNOWLEDGMENTS

The authors gratefully acknowledge the German Government, through the German Academic Exchange Service – DAAD (Deutscher Akademischer Austauschdienst) and the Brazilian Government, through the National Council for Scientific and Technological

Development – CNPq, for their financial support, which enabled the purchase of the sound level meters and software used in this study.

REFERENCES

Astolfi, A., Corrado, V., Griginis, A. Comparison between measured and calculated parameters for the acoustical characterization of small classrooms, *Applied Acoustics*, 2008, 69, 966-976.

Eyring, C.F. Reverberation time in "dead" rooms. *Journal of the Acoustical Society of America*, 1930, 1, 217-241.

Fitzroy, D. Reverberation formula which seems to be more accurate with nonuniform distribution of absorption. *Journal of the Acoustical Society of America*, 1959, 7, 893-897.

Fasold, W., Veres, E. Schallschutz + Raum Akustik in der Praxis. Verlag für Bauwesen – Berlin, 1998. (in German).

Frangos A. School Districts Spend to Ensure Good Acoustics. *Wall Street Journal* Jul 30, pg B1. New York, N.Y. 2003.

Hodgson, M. Rating, Ranking, and Understanding Acoustical Quality in University classrooms. The Journal of the Acoustical Society of America, Melville, 112(2), 568-575.

Harris, C. M. Noise control in building. New York: MacGraw-Hill, Inc., 1998.

Hodgson, M., Case-study evaluations of the acoustical designs of renovated university classrooms. *Applied Acoustics*, 2003, 65, 69-89.

Hohmann, R., Setzer, M.J., Wehling, M. Bauphysikalische formeln und tabellen. Editora: Werner Verlag, 4th edition. 2004. (in German).

International Electrotechnical Commission. IEC 60268- 16: Sound system equipment- Part 16: Objective rating of speech intelligibility by speech transmission index. Switzerland, 2003.

International Organization for Standardization. ISO 3382-1: Acoustics — Measurement of room acoustic parameters. Part 1: Performance spaces. Switzerland, 2009.

International Organization for Standardization. ISO 3382-2: Acoustics — Measurement of room acoustic parameters. Part 2: Reverberation time in ordinary rooms. Switzerland, 2008.

Jenisch, R., Klopfer, H., Freymuth, H., Petzold, K. Lehrbuch der bauphysic. Stuttgart B. G. Teubner Gmbh, 5th ed., 2002. (in German).

Lubman, D.; Sutherland, L. C. Good Classroom Acoustics in a Good Investment. *Classroom Acoustics*, p. 1-2, 2003.

Long, M. Architectural Acoustics. Academic Press, 2006.

Kang J. Numerical modeling of speech intelligibility in dining spaces. *Applied Acoustics*, 2002, 63: 1315-1333.

Millington, G. A modified formula for reverberation. *Journal of the Acoustical Society of America*, 1932, 4, 69-82.

NBR 12179 - Associação Brasileira de Normas Técnicas. NBR 12179: *Tratamento acústico em recintos fechados: Procedimento* [*Brazilian Association of Technical Standards. NBR*

12179: Acoustic treatment of indoor spaces: Procedure] Rio de Janeiro, 1992 (in Portuguese).

Neubauer, R. Classroom acoustics – Do existing reverberation time formulae provide reliable values? In: *Proceedings of 17th International Congress on Acoustics*. Rome. 2000.

ODEON Software Handbook - Version 9.0.

Passero C., Zannin P.H.T. Statistical comparison of reverberation times measured by the integrated impulse response and interrupted noise methods, computationally simulated with ODEON software, and calculated by Sabine, Eyring and Arau-Puchades formulas. *Applied Acoustics*, 2010, 71, 1204-1210.

Sette, W. J. A new reverberation time formula. *Journal of the Acoustical Society of America*, 1933, 4, 193-210.

World Health Organization – WHO. *Noise in schools*. Geneva, 2001.

Zannin, P.H.T., Passero, C.R.M., de Sant'Ana, D.Q. Acoustic Design of Enclosed Spaces. In: Ergonomics – Design, Integration and Implementation. Nova Science Publishers, Inc., 2008.

Zannin, P.H.T., Zwirtes, DPZ. Evaluation of the acoustic performance of classrooms in public schools. *Applied Acoustics*, 2009, 70, 626–635.

Zannin, P.H.T., Passero, C.R.M., de Sant'Ana, D.Q., Bunn, F., Fiedler, P.E.K., Ferreira, A.M.C. Classroom Acoustics: Measurements, Simulations and Applications. In: Classrooms – Management, Effectiveness and Challenges. Nova Science Publishers, Inc., 2012.

In: Noise and Ergonomics in the Workplace
Editor: Paulo Henrique Trombetta Zannin

ISBN: 978-1-62417-573-2

Chapter 5

ERGONOMICS AND ARCHITECTURE: INTEGRATION IN THE PROCESS OF CREATING WORKSPACES

Julia Issy Abrahão[1,*] and Claudia Bartolo Patterson[2]

[1]Department of Production Engineering,
Polytechnic School of the University of São Paulo, Brazil
[2]Office of Construction Management, Federal Justice Council,
Brazil

ABSTRACT

Integrating the ergonomic perspective into the scope of architectural design not only determines the necessary limits but also enables different solutions to be devised for the project that satisfy the specificities of the organization. Architectural programming is the instrument on which designers base their decisions in the various phases of architectural design. Understanding it as resulting from the investigation of the organization's needs and available resources and aligned with its objectives and intentions enables one to expand the possibilities of solutions to meet project requirements. Based on the assumption of interdisciplinarity, the areas of ergonomics and architecture addressed in this chapter are concatenated in a complementary manner to present a model of reference aimed at supporting designers in the development of architectural programs. The model aims to provide parameters of dimensional and spatial organization, using as reference the requirements of the activities to be performed in the space to be designed, thus assisting architects in the conceptual process.

1. INTRODUCTION

The non-correspondence commonly found between the physical spaces and intended use of corporate buildings suggests the difficulties that designers face in producing architectural designs that meet the qualitative and quantitative requirements of the needs of workers and

* E-mail: julia.abrahao@gmail.com.

the organization. Adjusting them to the project's intended purpose requires consistency of the architectural program with the working reality under consideration.

Identifying the dynamics of work situations and the interrelationships between workers and the organization promotes an understanding of the workspace as a system. Thus, its elements should not be viewed independently of their interrelationships. Data about the physical, anthropometric, physiological, social, cultural and psychological characteristics of the population of workers, as well as information about the work organization, prescribed tasks, functional hierarchy, working times, objectives, and instruments of production comprise the aspects that distinguish and singularize each work situation.

These interrelationships are manifested through the work activity and are one of the main bases for understanding and defining the goals of workers in their production commitment.

Different activity-related concepts are contradictory in amplitude and in the diversity of applications of the term in several areas of knowledge. Considering activity as the synthesis of the various aspects that determine it in the labor system, it is defined as the structuring (Abrahão, Sznelwar, Silvino, Sarmet and Pine, 2009), integrating and articulating (Béguin, 2007a) element of the interrelationships of this system. Similarly, activity is inserted within a given socio-technical context, which provides both the necessary resources for its implementation and its limitations.

The spatial and material conditions for the execution of work are factors that restrict or dynamize the interrelationships of the labor system, reinforcing the interdependence of each of the elements of the system (Abrahão et al. 2009). The identification of the nature of existing interrelationships intermediated by the work activity as a guiding principle of transformative actions and work concepts provides multiple solutions to problems that arise during the process of architectural conceptualization.

Based on the assumption of interdisciplinarity, this chapter presents a complementary combination of knowledge produced by ergonomics and architecture in order to define parameters, supported by methodological procedures, for use by designers in the development of architectural programs.

The proposed model aims to provide parameters of dimensional and spatial organization, supported by methodological procedures resulting from empirical studies, to underpin three architectural projects for the construction of corporate buildings (Abrahão, 1999; Abrahão and Patterson, 2005; Patterson, 2011; Patterson and Abrahão, 2009, 2011), using as reference the requirements of the activities to be performed in the space to be designed, thus assisting architects in the conceptual process.

2. Activity-Centered Ergonomics

At the beginning of its history in the mid-20th century, studies on ergonomics were applied essentially to the diagnosis of working conditions, especially in industrial environments. The current known as Human Factors emerged from the concern to achieve better performance conditions for human functioning in the design or transformation of human-machine systems.

European scholars, particularly in France and Belgium, directed their investigations to the study of an overall vision of labor conditions. The French approach to ergonomics, also known as activity-centered ergonomics, stands out for its approach and application, which differ from the Anglo-Saxon approach. The shift of emphasis in the study of the technical elements of the analysis of work in laboratory environments to real work situations has provided new opportunities for ergonomic actions. The establishment of the premises of study of the real work situation, of the interdisciplinary nature, and the possibility of effective participation of the actors in the work situation (Abrahão et al., 2009), have contributed to the development of knowledge and research into activity-centered ergonomics.

This approach goes beyond the notion that activity refers only to movements and gestures by including the subjective dimension of job execution (Abrahão et al., 2009). The decision-making and strategies used by workers to solve problems, and their interaction experiences in the context of real work situations (Wilson 2000) are a source of reviews and new investments in their role within the labor system (Béguin, 2007b). The composition of the dimensions that contextualize each work situation reveals the variability and the dynamics of the interrelationships among its elements.

Variability has to do with the result of intrinsic characteristics of workers and, at the same time, of work organization, and is susceptible to the influence of the dimensions of the production context on the interactions among the elements that make up a work situation. Rather than representing a constraint for ergonomic analysis, its knowledge represents the approximation to the working reality and the possibility of new projects for the solution of problems and requirements. The different dimensions that contribute to variability reveal the singular and unique characteristic of each working conditions (Abrahão, 2000).

3. Ergonomics and the Process of Conceptualization

The meaning of conceptualization used herein refers to the work of creation and design. This definition enables one to understand Béguin (2007a), who states that "conceiving is to pursue an intention" (p. 317). The act of conceiving, therefore, is a process that involves an objective, the direction to be followed, and the transformation of this objective into reality.

In addressing the role of the ergonomist in conceptualization processes, the "paradox of design ergonomics" of Theureau and Pinsky (1984, cited by Daniellou, 2007) deals with the impossibility of conceiving designs for the transformation of an observed activity, in that the transformation of working environments implies the transformation of the activity itself. This contradiction is broached by François Daniellou in addressing the "probable future activity," in which conceptual possibilities provide opportunities for different forms of future activities.

Designing is, in this sense, to consider the variability of work situations for a future population that is not always known. The "probable future activity" approach sees conceptual possibilities as opportunities for different forms of future activities. Thus, "reference situations" (Daniellou, 2007) become a powerful tool for the investigation of the future activity. Far from representing a model, they allow for the expansion of possible design options by different observable "sources of variability" (p. 308).

Understanding the activity contributes to the definition of parameters of reference of the future activity and its spatial requirements, translated by architectural programming as a support that enables the design. An attitude of openness to new proposals and the role of the ergonomist as an articulator of different views of reality contribute to his participation in different phases of the project (Martin, 2007). Anticipation of the future activity requires the early contribution of the ergonomist in the design process (Béguin, 2007a) and reveals the importance of ergonomic action in the process of architectural programming (Martin, 2007).

4. Architectural Conceptualization

The interaction between art and science finds, in the intent and purpose of the project, the binding element that transforms the building into architecture, and the physical space into architectural space. Thus, architecture can be understood as an open system subject to the influence of the environment and of the social, cultural, technical and scientific settings of a corresponding era (Santos, 2008).

Architectural design is the graphic record that guides the construction of a building. This design contains technical guidelines for execution and definitions relating to concrete architectural elements. The design also expresses the compositional elements, whose proportions, concepts and principles reflect the actions and the representations of intentions and needs of present and future users of the building (Martinez, 2000).

Despite the connection among elements that participate in the system, the connection between the intent and purpose and the conceived architectural space materialized by the building it is not always clearly identifiable. In this sense, architectural programming emerges as one of the guiding and organizing elements of the architectural concept. It is the instrument that provides the link between the expectation and the concretization of the desired space.

For the model proposed here, the process of architectural programming is based on three guiding principles: (a) Statement of the design problem, to which are associated the users' needs and the organization's resources; (b) the establishment of goals concerning the purposes and functions of the design; and (c) the adjustment between them (Figure 1).

Throughout the recent historical course of architectural programming, different methodological approaches resulting from studies and from professional practice have been published and have contributed to the development of its systematization.

However, the creation of architectural programming models as a reflection of social, economic and cultural conditioning factors does not mean an evolutionary sequence in which new paradigms replace the existing ones. Their application is also a choice made to better tailor the programming process to different project situations and requirements.

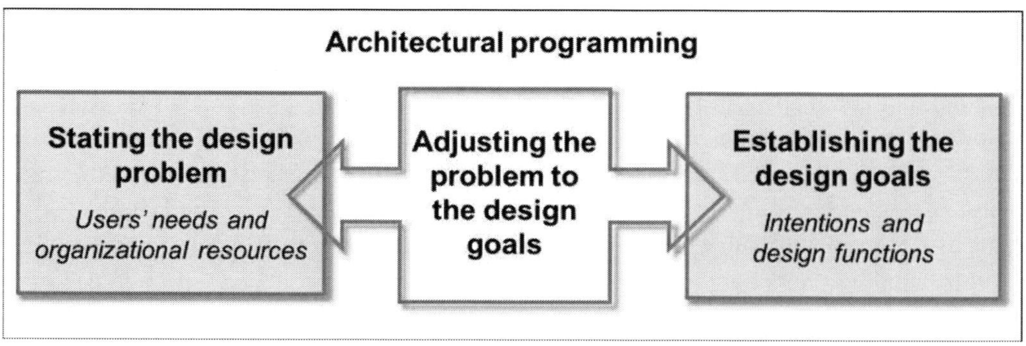

Figure 1. Objectives of the architectural programming process (Patterson, 2011).

The process should incorporate the requirements, context and promoting or limiting factors for the development of the project.

The methods employed in architectural programming seek to identify the singularity of each project (Cherry, 1999) and translate it into architectural elements that make up the design. In architectural programming, knowledge of the existing situation is therefore essential to strengthen its predictive character (Blyth and Worthington, 2007; Zeisel, 2006)

5. Ergonomics and Architecture

Ergonomics and architecture come together through planning and the investigation of activities. Through ergonomics, it is possible to understand human activities and their design requirements. The elements that enable activities to be performed can be provided through architecture.

The predominantly analytical nature of the architectural programming process finds in the ergonomic approach knowledge about how people work, the strategies used to perform their tasks, interpersonal communications, and the requirements for the execution of their work activities (Patterson and Abrahão, 2009, 2011). The complementary qualities of ergonomics and architecture thus contribute toward the construction of the architectural design, from the programming phase up to the evaluation of the workstation.

Analogously to ergonomic action, the construction of the architectural program has its own criteria and procedures. Its objective is to collect data that will support the architectural design. In the case of a design destined for a workspace, this is preceded by an understanding of the different points of view and the objectives, desires, intentions and limitations of the workers and the organization. Its product results from a global vision of the system, taking into account the quantitative and qualitative findings of the collected data. Both quantitative and qualitative variations occur in the spatial elements of particular contexts (Santos, 2008), revealing the diversity and variability of work situations (Abrahão et al., 2009).

Studies aimed at improving the prediction of workers' actions as a function of their workspace indicate the importance of approximating reality through knowledge of the activity. Knowing the context of the activities directly related with the performance of the task, as well as so-called mediating activities such as making copies, answering the telephone, and the worker's physical and psychological needs, such as drinking water or coffee, going to the toilet, or taking a break (Tabak and de Vries, 2010), may affect design-related decisions about the occupation and use of the workspace.

The ergonomist's contribution to the architectural design is often related to ergonomic requirements of environmental comfort in terms of lighting, ventilation, acoustics and furniture (Cherry, 1999; Martin, 2007). However, the involvement of the ergonomist has increased considerably in all the phases of the architectural project, including the preparation of the program.

The proposed model, in addition to the requirements of the physical conditions of the environment and the furniture, also considers the elements related to safety and accessibility in different situations, the ease of work flow, the cooperation and interaction among workers and users, the correspondence between the organizational structure and the physical context of the workspace, and the need for privacy (Margaritis and Marmaras, 2007). The actions of analysis, synthesis and evaluation are revealed in the feedback from the different phases of the process, in which "certain components, restrictions or specification can only emerge along the process, [...] and in the interactions between actors of the conceptual process and users" (Martin, 2007, p. 359).

To consider the assumption that an organizational reality is subject to change enables one to question the validity of the architectural program as a representation of a single given moment, considering the entire set of the architectural concept (Patterson, 2011). However, architectural programming, similarly to ergonomic action, must be understood as a dynamic process susceptible to adjustments and modifications. Although the concept of functional flexibility is introduced right at the beginning of programming, the more knowledge is available about the present moment, the better the chances for success of the project as a positive response to the demand problem (Patterson and Abrahão, 2011).

An attitude of openness to new proposals and the role of the ergonomist as a mediator and articulator of different views of reality contribute to his participation in different phases of the project (Martin, 2007). The application of analytical methods and the resulting production of knowledge about activity in its structuring role of work situations enable the ergonomist to identify typical situations of action that can be translated to the spatial conception of a future activity (Garrigou, Daniellou, Carballeda, and Ruaud, 1995), through scaling and the determination of the guidelines for the architectural project.

Typical action situations in a given work system represent typical situations of occupation, use and arrangement of the space of this system (Patterson and Abrahão, 2011). The correspondence between the analysis of action situations and space-related situations therefore requires the communication and interaction of the ergonomist's knowledge and investigation techniques, as well as the architect's technical mastery of the design process.

6. The Model

The purpose of applying the different phases of ergonomic work analysis (EWA) in the three empirical studies used as reference for the proposed model was to add a differential to the construction of the traditional architectural program that would integrate not only the tasks and the organizational structure of corporations, but above all, the activities they perform, aiming to minimize potential problems when the space becomes occupied (Abrahão and Patterson, 2005; Patterson, 2011; Patterson and Abrahão, 2009, 2011).

The strictness of the internal rules of the organizations that participated in this study, however, did not allow for the architectural programming to follow along with the different phases of the design, which would have enabled its product – the architectural program, to be the main guiding tool of the design possibilities. This condition inevitably required that the architectural program contain specific information at a sufficient level of detail that would render the future procurement process feasible. The inclusion of the parameters of the present situation, and particularly the parameters of future reference, contributed not only to the projects that might eventually be developed by the company's technical staff but also to the technical basis for projects that may be contracted in future. Thus, the model adopted was the one in which the need to separate architectural programming from the design process (Cherry, 1999, Peña and Parshall, 2001) incorporates the procedures for internal construction procurement processes, standardized by the organizations that served as reference for the study (Figure 2).

The contexts of the project and time constraints did not allow for an ergonomic analysis along traditional lines. The corporations' internal rules and administrative procedures imposed limits on the decisions concerning the different phases of preparation of the architectural project. Therefore, the procedures and tools were reformulated while maintaining the basic principles of EWA, ensuring worker participation in the process of conception.

7. The Method

The purpose in defining the elements of reference for a methodological procedure to support architects during the architectural programming process using the EWA approach enabled the parameters and goals of the design to be aligned with the reality of work allied to the working needs.

The method was constructed by means of adjustments, in a feedback process referenced by the singularity of the data obtained in each phase. Using the activity as the guideline of the investigation, the qualitative and quantitative determinants of the elements and of the interactions of systems of actions and of objects were taken into consideration without, however, losing the overall vision necessary to ensure the coherence of the space to be designed. The proposal of the method adopted for architectural programming and EWA proved to be essential to diminish the gap between the architectural programming and the activities of the work situation (Patterson, 2011; Patterson and Abraham, 2009, 2011).

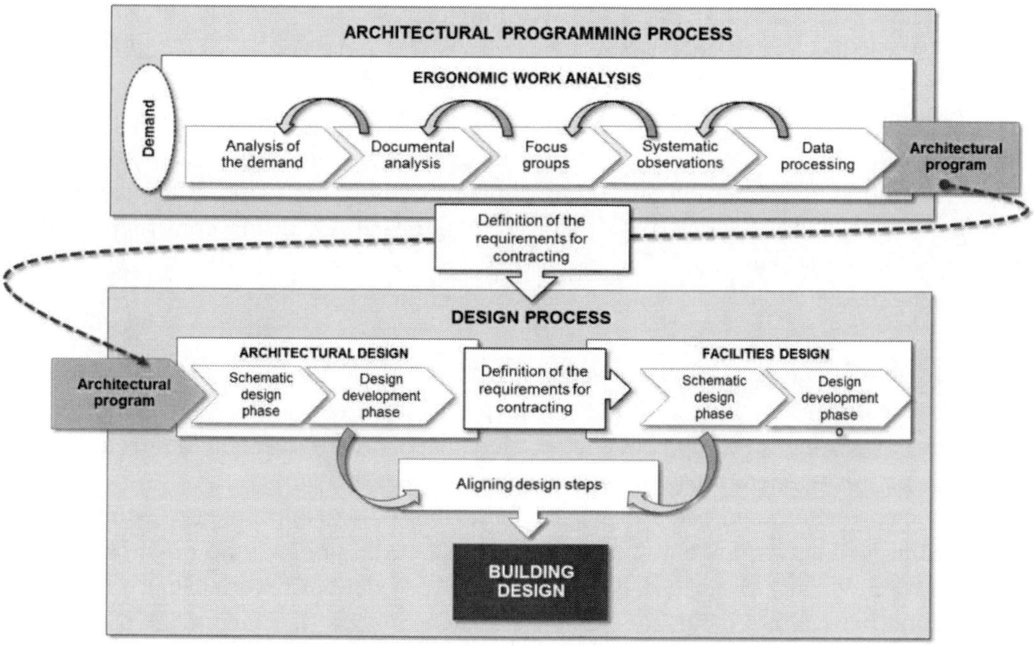

Figure 2. Architectural programming process based on EWA (Patterson, 2011; Patterson and Abrahão, 2011).

The phases of the method were organized based on the answers to the following questions:

- How can the nature of the activities be identified?
- How can quantitative and qualitative parameters be established that meet the needs of workers and the organization, as well as the design requirements, by means of the architectural program?
- How can the architectural concept of the socio-technical conditioners be brought in line with the reality of work?

Case studies reported in the literature as a reference for the methodological approaches used in the architectural programming process reveal adjustments and adaptations of these methods to each particular situation (Blyth and Worthington, 2007; Cherry, 1999; Zeisel, 2006). The architectural program resulting from the programming process does not correspond to a design solution, but to guidelines that lead to design solutions.

An analysis of models proposed for the architectural programming process, such as those of Peña and Parshall (2001), Edith Cherry (1999), Donna Duerk (1993) and Blyth and Worthington (2007), revealed the variety of parameters used. Using these models as reference and integrating the context, constraints, and characteristics of the design requirements, the following assumptions were defined for the construction of the model adopted here:

1. Preceding programming process dissociated from the architectural design (Cherry, 1999).
2. User participation in data collection and validation (Blyth and Worthington, 2007; Cherry, 1999; Peña and Parshall, 2001), in a collaborative process that considers different points of view (Blyth and Worthington, 2007).
3. Contextualization of the design problem (Blyth and Worthington, 2007; Cherry, 1999).
4. Prioritization of abstract thinking in detriment to predefined design solutions (Cherry, 1999).
5. Description of the current and future situations within the scope of the data, adding the predictability factor (Blyth and Worthington, 2007) in the construction of the architectural program.
6. Cyclic model of analysis, synthesis and evaluation (Duerk, 1993).

The categories were established considering the physical conditions of the workspace resulting from existing or future activities. The space was dimensioned as a function of the nature of the task, furniture, equipment and the elements of environmental comfort in terms of lighting, ventilation, temperature and noise (Abrahão et al., 2009).

The structural nature and performance of the activity also contributed to demarcate the aspects relating to interactions between the workers and the units of the organization. The purpose of these elements of analysis was to identify the relationship of spatial proximity or communication, providing input for design decisions about the functional aspects and distribution of areas.

Lastly, taking the actual activity as unit of analysis enabled a broader and deeper understanding of the interactions that take place between the systems of the work situation and of the workspace.

7.1. Construction and Definition of the Procedures and Instruments

Given the context of the demand in the application of the model, it was decided to set up a multidisciplinary team comprising professionals from the areas of ergonomics, architecture, engineering, and administration. The team's activities were coordinated by the ergonomist.

The phases of ergonomic analysis are planned with a view to enriching the architectural programming. From this perspective, data originating from the ergonomic methodology are integrated, at different moments, with the usual data of the architectural design process for the definition of collection procedures. These procedures allow for the feedback of data that are part of the elements of the activity that would otherwise not be considered in the design, based on an approach restricted to a single area of knowledge.

The first data that emerged from the documental analysis of the organizations raised various types of questions that were clarified by the managers through a collective interview. The consolidation of the resulting data allowed for the inclusion of the functional structures that may eventually be integrated into the existing organizational structure. The analysis of the probable changes in assignment, the increase in the number of workers, and the creation of new functional units served as the basis for the guidelines of the architectural program.

The scope, the complexity, the large number of employees, and the short period of time available to determine the demand were instrumental factors in reflecting about the procedures and instruments to be adopted in the collection of data, while simultaneously remaining true to the guiding principles of the ergonomic action. Considering the time constraints, a traditional ergonomic analysis would preclude the execution of the architectural program as desired. Accordingly, the first challenge was to determine the best way to integrate the user of the workspace in order to define and build the parameters that would constitute the guiding principle of the architectural program. In this sense, the work designed based on the worker's perception would become the initial cornerstone of analysis and would underpin our understanding of the worker's demands.

The process was initiated with an instrument sent to each worker, asking him to describe the equipment, special facilities and furniture used in his sector, as well as the ones that would be required for the proper performance of his activities. The data thus obtained were categorized and grouped according to their frequency of occurrence.

Later, it was decided to set up focus groups to collect data from each unit in the organization. The objective was to understand the collective perception about the work spaces and the functional requirements of each unit in particular. The results obtained through the individual instrument were presented to the group. The members of the multidisciplinary team participated as mediators and facilitators in the discussion sessions with focus groups. Guiding questions were linked to the four categories established from the initial collected data, as indicated in Table 1.

The data collected through this procedure were typed into a computer and examined by the group participants with the help of data display equipment.

Table 1. Correspondence between the categories of analysis and the questions asked to obtain data from the focus groups (Abrahão and Patterson, 2005)

Categories	Questions
Activity	What types of activities are performed?
	Is there external customer service, internal customer service, or both? In what situations is the service provided?
Physical conditions of the workspace	What furniture and/or appliances are needed to perform the above described activities?
Improvement suggestions	How many people work in your sector and what positions do they occupy?
	Does the organizational chart of your sector reflect the reality of its work? If not, what suggestions would you offer to render it adequate for the activities performed in your sector?
	What are your expectations regarding the design of the new building in terms of improving the quality of life in your work environment?
Interactions among sectors	Which sectors should be close to yours?
	With which sectors do you need to exchange information in order to perform your tasks?
	From which sectors do you receive job requests?
	When jobs are concluded, to which sectors are they passed on?

The sessions started with presentations by the participants and the team that headed the work. The team then presented the rules for the meeting.

At the end of the session, the records that had been displayed were read by everyone present. If the participants of the focus group perceived that important points had not been discussed, these were immediately included. After the reading, the attending workers validated the information before the session was concluded. At the request of the managers of the organizations, the collected records were printed and sent to them for a subsequent second validation of the data.

The data collected through the focus groups provided input for the qualitative parameters of the activity and of the physical workspace. After these data were processed, it was found that they needed refinement in order to integrate the work dynamics and thus validate them in the workspace while carrying out the activities in real time. We sought a unit of the organization that would enable the validation of collected data, and that would consider the requirements of activities representative of the work performed. In this unit, the data were collected through systematic observations of the workspace and photographic records, aiming to complement the quantitative data necessary for the architectural program.

7.2. Categories of Analysis

The determination of categories for data collection was based on the activities and future projections for the organization as a function of the project, associated with the units of analysis concerning the quantitative aspects of the dimensioning of physical workspace and the aspects of social and organizational interactions.

Similarly, to determine the categories, integrating the ergonomic viewpoint in which the activity is the main point of the investigation process allowed for aspects of the physical conditions of the workspace to result from present or future activities. The dimensioning of the space took into consideration the activity as a function of the nature of the task, furniture, equipment and the circumstances of the elements of environmental comfort in terms of lighting, ventilation, temperature etc. (Abrahão et al., 2009).

The structural and at the same time articulatory nature of the activity also contributed to the definition of aspects relating to the interactions between the workers and the units of the organization. These elements of analysis provided information about the relationship of spatial proximity or of communication, offering data for design-related decisions with regard to functional aspects and the distribution of areas.

Finally, considering the activity itself as a unit of analysis enabled a broader and deeper understanding of how the systems of the work situation and the workspace interact.

Knowledge about the activities for architectural programming helped shed light on the conditions of space and communication, providing the input data for the architectural design. For example, based on the architectural program, the designer was able to determine which activities required their own space and which activities could have their space shared with others.

The collection of data from the focus groups was established based on the results obtained from the first individual instrument. The questions were designed to determine: (a) the general data of the organizational unit; (b) the characteristics of the activities performed by the workers; (c) a description of the procedures and work routines as well as the strategies for carrying out the assigned tasks; (d) a description of the furniture and equipment/appliances necessary for the activities; (e) the interrelationship between the functional units and their relationship of proximity, in order to identify cooperation in the work, and the best location in the building; (f) the type of service provided to the internal public or to external visitors, in what situations it occurs and what equipment and furniture are required; and (g) the workers' expectations about the new building, including their representations about the organization, the work environment and the social relations among the workers.

The data were collated on spreadsheets, on which the focus groups from which they originated were identified.

To facilitate the analytical procedures, subcategories were established to group and summarize data of the same nature. The frequency of reference to each item of information was recorded and added up at the end of the analysis. This procedure enabled the recognition of recurrent data in the representations of the participants of the groups. Finally, the construction of a table listing the frequency of reports arranged in decreasing order facilitated the visualization of their elements for analysis.

In the category "Improvement suggestions," in addition to processing the data as described, data on the number of workers and the organizational structure were listed on a spreadsheet containing the situation at that time and the expected future situation.

As for the category "Interactions among sectors," the construction of a comparative chart containing the data of each focus group in relation to the other units of reference of the organization enabled the visualization of consistencies and inconsistencies in the depictions of the workers, and served as feedback for new procedures.

The correspondence achieved by cross-referencing the data of the categories of analysis by means of technical project elements transformed subjective information into architectural determinants for the workspace. The variable of the organizational size of the functional hierarchical structure was added to the modules of areas, represented graphically by schematic designs, and consolidated the elements identified in the categories of analysis, together with the quantitative elements of reference for the architectural programming.

The data obtained through the systematic observations were processed similarly to the data of the focus groups. The activities and their frequency of occurrence within the period of observation were identified and recorded in tables to facilitate their recognition and comparison of the data for analysis. The analysis of photographic records in comparison with the data obtained enabled the recognition of the component elements in the representations of the employees.

The data of the different phases were cross-referenced, formalized by means of tables, qualified and presented in architectural programming format, which determined the conditions of the architectural designs for the new headquarter buildings of the corporations.

8. Contributions of the Focus Groups

The decision to set up focus groups led to the effective participation of workers in the collection of data, integrating their representations and in-depth knowledge about their activities. The participants' interaction with respect to the proposed issues enabled the plurality of representations to be understood collectively, seeking consensus and confirming the influence of the social and collaborative nature of the discussions. The meetings allowed for information exchanges about the structuring activities of the work system and about the systems of objects and actions in the workspace, existing and desired, brought up by the employees themselves.

The diversity of points of view that reflected these realities was essential in understanding the interrelationships and interactions of the different organizational units, and enriched the representations as data of the requirements for the architectural program.

The analysis of the categories "Activities" and "Physical conditions of the workspace" were identified in the different actions reported in response to the questions corresponding to them. The data were marked according to the participating unit and their frequency in relation to the number of times they were mentioned by the focus groups, and are presented in tables and arranged in descending order by the number of records in the focus groups.

The lack of quantitative and qualitative data obtained in the initial interview with the managers resulted in dividing the analysis of the category "Improvement suggestions" into two distinct processes. The first dealt with the increase in hired personnel and a possible functional restructuring of the organization. The quantitative data were classified in terms of the number of workers per unit, their positions in the situation existing at the time of data collection by the focus groups, and the number of workers probably needed to align the organizational structure with the work reality of the respective sectors. In the second process, which is the main object of interest for the proposed model, the workers' expectations about improvements in the quality of the work environment were analyzed according to the procedure adopted for the categories "Activities" and "Physical conditions of the workspace," incorporating data with characteristics similar to those of the subcategories of analysis. In the research for the architectural programming, the inclusion of the category "Interactions among sectors" provided a systemic vision of the work situation structured on the activity to be the main guideline for an understanding of the need for communication and physical proximity among different units of the organization.

One of the key elements in the architectural programming proved to be communication, enabling the constructed space to approach the work reality of the employees in the collective relationship of their work. This precluded the usual practice of identifying proximity by means of the vertical relationships of subordination between the units in the functional organization chart.

8.1. Analysis of the Category "Activities"

For a better examination of the similarity of the characteristics described by the workers in their representations, the collected data were grouped by the focus groups into ten subcategories of activities:

1. Documental: actions involving planning, preparation, analysis and production of documents.
2. Virtual: work performed using informatics tools.
3. Meetings.
4. Customer service: Includes face-to-face, telephone and electronic mailing service.
5. Administrative: involves work routines.
6. Support: provides supporting or operational material for the execution of other activities.
7. Specialized techniques: refers to work whose requirements of knowledge in any specific technical area differentiate it from other activities.
8. Technical support: differs from support activities due to its technical nature.
9. Group work.
10. Prerequisites: refers to requirements understood as necessary to carry out a task.

Grouping the activities into subcategories enabled a more precise identification of the characteristics and nature of the activities reported by the focus groups.

Thus, the synthesis of activities, in terms of their essential aspects, reduced the categorized activities, as shown in Table 2. The descriptions of the activities involved in the subcategories of documental, administrative and customer service varied significantly in the reports of the focus groups.

The minor interventions made to synthesize the subcategory of specialized technique activities, and the absence of modifications in the activities related to the technical support and virtual subcategories indicated the specificity of the work performed, principally due to the requirement of particular qualifications for their execution. The data analysis enabled the nature and singularity of the execution of the activities to be distinguished by quantitatively identifying the most frequent ones and the ones recorded only rarely by the focus groups.

Table 2. Subcategories of analysis of the category "Activities" (Patterson, 2011)

Category "Activities"	
Subcategories	Description
Documental	Preparation and analysis of texts, processes and documents.
	Preparation, coordination and follow-up of projects.
Virtual	Issuance and management of contracts
	Systems management
	Database administration
	Work via systems
Meetings	Internal meetings
	Meetings with managers
	Meetings with commissions
	Meetings with shareholders
	External meetings
	Video conference meetings
Customer service	External customer service
	Internal customer service
	Advice to companies via e-mail or telephone
	Assistance to authorities
	Over-the-counter customer service
	Press relations
Administrative	Routine activities (administrative, routine tasks, receipt and distribution of documents from other areas).
	Filing of processes and documents
	Dispatch of materials and documents
	Registration of documents
	Media archiving
	Control and custody of fixed assets and warehouse stock
Support	Leading commissions
	Setup of material for meetings (folders)
	Leading meetings
	Correction of graphic material
	Setup of material for events (folders)
	Reprographics
	Arrangement and monitoring of events
	Arrangements for the departure and arrival of executives and authorities
Specialized techniques	Preparation and analysis of architectural and engineering projects
	Production of journalistic material (recording, filming, editing, etc.)
	Communication services (telephony and radio communication)
	Bidding
	Monitoring of distance courses
	Call Center
	Culling filed documents for discard

Category "Activities"	
	Collection and processing of material (books, articles) for the library collection
Technical support	Management and maintenance of equipment and installations
	Management of information technology facilities
	Selection of books and media material
	Equipment distribution and installation
	Equipment evaluation and testing
Group work	Group work with employees
	Group work with external public
	Interviews with users
Prerequisites	Work requires concentration
	Work requires communication among employees
	Work requires creativity

Based on these results, it was possible to infer that the requirement of specific knowledge on the part of employees for the execution of their tasks, as well as the necessary appliances, equipment, and furniture, is an essential condition for the configuration of their workspace.

8.2. Analysis of the Category "Physical Conditions of the Workspace"

The classified data of the category relating to the physical conditions of the workspace was arranged into the following subcategories:

1. Furniture.
2. Equipment: elements involved in carrying out the activities whose operation requires building infrastructure facilities such as power supply, logical and telephony cabling.
3. Special equipment: destined specifically for a given activity due to the requirement of special installations, or space for their installation.
4. Accessories: elements complementary to the furniture required to carry out the work.
5. Environments with special installations: spaces whose singular activity must, among other elements, meet the ergonomic requirements of environmental comfort, specific furniture, location suitable for the activity, and a separate design.
6. Support environments: spaces destined for support activities or services.
7. Design premises: general guidelines for the architectural design, destined for a particular activity or workspace.

The criterion for synthesizing the data into subcategories was based on their similarity or their relevance to a larger cluster. For example, eight different shapes were recorded for the cabinets described in subcategory "Furniture," ranging from simply "cabinets" to a detailed specification as "closed cabinets for office material" or "cabinets for storing software and documents."

This procedure led to a significant reduction of the records initially prepared by the focus groups, facilitating their subsequent quantitative and qualitative analysis, similarly as the analysis made for the category "Activities."

The parameter of the focal question of the category "Physical conditions of the workspace" was furniture and the equipment required to carry out the activities listed in the previous category.

The analysis of the data revealed that the employees' representations of this category are extensive and point to the workspace as a reference for the investigated data. Therefore, in addition to the subcategories relating to furniture and equipment, there are data that specify the workspace and the activity to be performed in it, as in the subcategories "environments with special installations" and "support environments."

Taken as a subcategory, however, the design premises identified in the representations of the focus groups are the expression of the employees' participation in creating the guidelines to be followed in the conception of the architectural design. Hence, aspects of lighting, thermal comfort, acoustic comfort, visual comfort, visibility, privacy and accessibility are characterized as part of the favorable conditions for the composition of the space to be designed.

However, the apparent contradictions found among the representations of the participants of focus groups about their activities and about the elements of the physical space reinforce the importance of the global vision and the technical and articulating knowledge of the ergonomist to ensure coherence among the different components of the work system.

The close correlation of the data of the category "Physical conditions of the workspace" with those of the category "Activities" was clearly identified in different subcategories.

From the ergonomic standpoint, the description of the component elements of the workspace system favored the configuration of the qualitative and quantitative determinants of the architectural program used by the designer, thus aligning the design with the work reality.

8.3. Analysis of the Category "Improvement Suggestions" with Emphasis on the Workspace

The subcategories identified by the focus groups and later through systematic observations concerning improvements of the work environment were:

1. Work environments: guidelines or the need for a set of conditions for a specific activity, involving material, organizational, cultural and other aspects.
2. Support environments: a set of conditions destined for support activities or for commercial and services structures complementary to the organizational structure.
3. Spaces for socialization: a set of conditions desired for a space for workers to meet and socialize.
4. Furniture: proposed guidelines for adapting the furniture to the nature of the activities.
5. Facilities and equipment: guidelines for the execution of activities or for adjusting the environmental conditions of the workspace.
6. Architectural elements: building materials, finishing materials, or parameters for decisions-making about the architectural design.
7. Ergonomic requirements of environmental comfort: these refer to the physiological and perceptual requirements for the performance of work.

8. Structure organizational: suggestions for the improvement of the functional organization.
9. Socioenvironmental requirements: actions aimed at contributing to social inclusion, economic viability and environmental preservation.

After synthesizing these data, it was found that, more than simply suggestions for improvement, the data obtained from the focus group represented different aspects of the interaction between the workers and the organization to which they belonged.

The analysis revealed that the category "Improvement suggestions" contains data that are very similar to those collected in the categories "Activities" and "Physical conditions of the workspace," indicating the tenuous boundary between desire and real need (Peña and Parshall, 2001). However, the intentions represent the future situation to be considered in the design, suggesting that the ergonomist take into account the indicators contained in the representations of the participants in the process. The subcategories "Work environments," "Furniture" and "Facilities and equipment" reflect need in the form of desire. Concealed by these representations is the nature of the activity as a determining factor of the qualitative variables considered in the programming process.

Although many reference data were provided by only three focus groups, their importance for the composition of the architectural program should not be overlooked. In his role as mediator of the different points of view about the work reality, the ergonomist considers, in his analysis, the typical activity situations and also the specificities of the situations under study.

8.4. Analysis of the Category "Interactions among Sectors"

To paraphrase Wilson (2000), a correct understanding of the interrelationships and interactions among different components of the organizational structure is part of the action of the ergonomist. The object of this action is to incorporate the data about the cooperation and proximity among the sectors of the organization as part of the investigation of the workers' activities and needs in the process of layout design.

The data collected from the focus groups resulted from the answers to the four questions proposed for the category "Interactions among sectors." The following subcategories were used as reference for the analysis: (a) no interaction among the sectors; (b) communication among units, by means of information exchange between sectors, without the requirement of physical proximity; (c) physical proximity among units as a factor that favors the execution of their activities; and (d) communication and physical proximity among the units.

An analysis of the number of references to interactions mentioned by the focus groups with regard to units revealed significant differences. A comparison of these data allows for a qualitative analysis of the discrepancies among the groups. The importance of this analysis lies in the correct understanding of the interrelationships and interactions among different sectors of the organization and the ratios of efficiency and effectiveness that they determine in the physical space. For architectural programming, these data are used to establish guidelines for the design in terms of the physical location of these sectors in order to promote or inhibit communications, according to need.

Inconsistencies were found in two directions: inconsistencies in the data of the focus groups separately in relation to the data from the other units of reference, and discrepancies in the data of the units of reference vis-à-vis the data of the focus groups.

Six discrepant situations were identified:

1. The focus group does not mention the interaction, but the units of reference require communication with the sector of this group.
2. The focus group does not mention the interaction, but the units of reference require communication with and physical proximity to the sector of this group.
3. The focus group requires communication allied to the need for physical proximity, but the units of reference mention only communication with the sector of this group.
4. The focus group requires communication, but the units of reference do not mention interaction with the focus group.
5. The focus group requires communication, but the units of reference require communication allied to the need for physical proximity with the focus group.
6. The focus group does not mention interaction, but the units of reference require physical proximity with the sector of this group.

For the model presented here, the inconsistencies found in items 2, 3, 5, and 6 prompt special interest in the model. The indication of interaction with the requirement of physical proximity has an impact on the quantitative and qualitative aspects incorporated into the architectural program. However, the fact that physical distance or difficulties in face-to-face communication may lead to negative interference and the breakdown of collective work necessary for the system's operation is less obvious than the interference of aspects concerning the incompatibility of contiguity between sectors as a function of the nature of their activities, such as, for example, the proximity of a sector whose activities are a source of noise to a work area that requires concentration.

Conclusion

Combining the ergonomic approach with the parameters and categories of analysis chosen for the architectural programming strengthened the design guidelines and promoted the link required to satisfy its prerequisites. Through its different instruments, the comparison of the data analyzed in the light of interdisciplinarity integrated the diversity of variables and the multiplicity of actions. Thus, the assumption of the participation of stakeholders of the work situation in the conception process is corroborated by the participation of users in both the collection and validation of data and in the intentions for the configuration of the guidelines for the architectural design.

Interest in the study of the real work situation emphasized the singularity of its design requirements and the need to contextualize it, contributing to its articulation with the future situation. This approximation to the work reality enabled the construction of the architectural program as a predictive tool for the development of projects.

The reference of the activity as the central organizing element of the workspace system enabled categories to be established, which were then added to the architectural program. The definition of area module based on the data collected through the representations of the workers contributed to the incorporation of qualitative elements in the dimensioning of the design. Lastly, systematic observations that demarcated the situation under study substantiated the validation of the data.

The ergonomic approach associated to the study through its different instruments allowed for the comparison of data sets, analyzed in the light of interdisciplinarity, integrating the diversity of variables and the multiplicity of actions. The distinctiveness of the approach of ergonomic workplace analysis, in terms of methodological procedures and concepts, enhances the interface between ergonomics and architecture in the development of architectural programs.

The methodological support of EWA helped to confirm the data that were representative of and closest to the work reality. Indeed, relying only on an examination of the data of the representations of the focus groups would not have ensured the successful characterization of the relationships of communication and physical proximity. From this analysis it can be inferred that cross-referencing the collected data reveals the need for validation using other instruments and techniques.

The conceptual and methodological interfacing of ergonomics and architecture in the model presented in this chapter indicates the possibility of filling in gaps in the architectural programming process. The expansion of this study through new investigations is expected to involve the use of informatics tools for graphic simulation and the creation of a database of area modules as a reference for architectural programming in corporate organizations.

REFERENCES

Abrahão, J. I. (1999). Configuração espacial de um prédio para postos de trabalhos informatizados [Spatial configuration of a building for informatics workstations] (Fundação de Seguridade Social). Unpublished manuscript.

Abrahão, J. I. (2000). Reestruturação produtiva e variabilidade do trabalho: Uma abordagem da ergonomia [Productive restructuring and work variability: An ergonomic approach]. *Psicologia: Teoria e Pesquisa,* 16 (1), pp. 49-54.

Abrahão, J. I., and Patterson, C. B. (2005). Programa de Atividades para a Nova Sede do Tribunal Superior Eleitoral [Architectural program for the new headquarters of the Superior Electoral Court] (Contrato TSE N. 52/2005 – Anexo I - Projeto Básico). Brasília: Tribunal Superior Eleitoral.

Abrahão, J. I., Sznelwar, L., Silvino, A. M. D., Sarmet, M. M., and Pinho, D. L. M. (2009). Introdução à Ergonomia: Da prática à teoria [Introduction to Ergonomics: From practice to theory] (a textbook). São Paulo: Blücher.

Béguin, P. (2007a). O ergonomista, ator da concepção [The ergonomist, actor of conceptualization] (G. M. J. Ingratta, M. Maffei, M. W. R. Sznelwar, M. A. de Oliveira, and A. A. Puntch, Trans.). In P. Falzon (Ed.), Ergonomia (pp. 318-330). São Paulo: Blücher.

Béguin, P. (2007b). Taking activity into account during the design process. @ctivités, 4 (2), pp. 115-121.

Blyth, A., and Worthington, J. (2007). Managing the brief for better design. Oxon: Spon. Press.

Cherry, E. (1999). Programming for design: from theory to practice. New York: Wiley.

Daniellou, F. (2007). A ergonomia na condução de projetos de concepção de sistemas de trabalho [Ergonomics as means for the conceptual design of work systems] (G. M. J. Ingratta, M. Maffei, M. W. R. Sznelwar, M. A. de Oliveira, and A. A. Puntch, Trans.). In P. Falzon (Ed.), Ergonomia (pp. 303-315). São Paulo: Blücher.

Duerk, D. P. (1993). Architectural programming: Information management for design. New York: Wiley.

Garrigou, A., Daniellou, F., Carballeda, G., and Ruaud, S. (1995). Activity analysis in participatory design and analysis of participatory design activity. *International Journal of Industrial Ergonomics*, 15, 311-327.

Margaritis, S., and Marmaras, N. (2007). Supporting the design of office layout meeting ergonomics requirements. *Applied Ergonomics,* 38, 781-790.

Martin, C. (2007). O ergonomista nos projetos arquitetônicos [The ergonomist in architectural designs] (G. M. J. Ingratta, M. Maffei, M. W. R. Sznelwar, M. A. de Oliveira, and A. A. Puntch, Trans.). In P. Falzon (Ed.), Ergonomia (pp. 357-369). São Paulo: São Paulo: Blücher.

Martinez, A. C. (2000). Ensaio sobre o projeto [Test on the project] (A. L. Spaltemberg, transl.). Brasília: UnB.

Patterson, C. B. (2011). Ergonomia e Arquitetura: Interfaces na elaboração de programas arquitetônicos [Ergonomics and Architecture: Interfaces in architectural programming]. (Master's thesis, Universidade de Brasília, Instituto de Psicologia, 2010). Repositório Institucional da Universidade de Brasília, 213. Retrieved March 20, 2011, from http://hdl.handle.net/10482/7154.

Patterson, C. B., and Abrahão, J. I. (2009). Real space: The common dimension of Ergonomics and Architecture. Proceedings of the 17th World Congress on Ergonomics, IEA2009 (CD-ROM), Beijing: International Ergonomics Association.

Patterson, C. B., and Abrahão, J. I. (2011). A programação arquitetônica sob a ótica da ergonomia: um estudo de caso no setor público [Architectural programming from the perspective of ergonomics: a case study in the public sector]. *Ambiente Construído*, 11 (3), 177-195.

Peña, W. M., and Parshall, S. A. (2001). Problem seeking: An architectural programming primer (4th ed.). New York: Wiley.

Santos, M. (2008). Espaço e método [Space and method] (5th ed.). São Paulo: Edusp.

Tabak, V., and Vries, B. de (2010). Methods for the prediction of intermediate activities by office occupants. *Building and Environment*, 45, 1366-1372.

Wilson, J. R. (2000). Fundamentals of ergonomics in theory and practice. *Applied Ergonomics,* 31, 557-567.

Zeisel, J. (2006). Inquiry by design: Environment / behavior / neuroscience in architecture, interiors, landscape and planning (Rev. ed.). New York: Norton.

In: Noise and Ergonomics in the Workplace
Editor: Paulo Henrique Trombetta Zannin

ISBN: 978-1-62417-573-2

Chapter 6

ERGONOMIC DESIGN

Carolina Reich Marcon Passero [*]
and Lucy Mara Silva Baú
Fisiotrab – Ergonomics, Health and Safety at Work, Brazil

ABSTRACT

Ergonomic design is the adaptation of working conditions to human limitations and abilities in the design phase of a new physical installation, a new work system, or new products or tools. Based on this concept, an analysis was made of the implementation of ergonomic design in an industrial unit, with a steam raising boiler as the object of study. The method employed was Ergonomic Analysis of Future Work, with the participation of multidisciplinary teams composed of operation, maintenance, and industrial safety technicians, ergonomists, designers and engineers. A flowchart was generated to analyze the requirements and, using Microsoft Office Access software, a tool was created to manage the database developed by the multidisciplinary team. During the 18 months of design analysis, 95 ergonomic requirements were identified. These requirements related mainly to the difficulty or impossibility of access to valves, instruments, equipment, electrical panels and viewers (68.5%) and of circulation in the industrial area (31.5%). All the recommendations for ergonomic improvements, which were discussed and accepted by the multidisciplinary team, were implemented in the field before the startup of operation of the boiler, at no additional financial cost to the company in which the boiler was installed.

1. DESIGN ERGONOMICS

Ergonomics is the scientific discipline concerned with the understanding of the interactions between humans and other elements of a system (International Ergonomics Association, 2000). Van Wely (1970) affirms that the purpose of ergonomics is to promote comfort, efficiency and health. Brazil's Regulatory Standard no. 17 (NR-17) – Ergonomics

[*] Email: cal.marcon@gmail.com.

(Ministry of Labor and Employment, 1978) establishes parameters that allow for the adaptation of working conditions to the psychophysiological characteristics of workers.

According to Iida (2005), design ergonomics involves the adjustment of work to human characteristics during the design of a product, machine, environment or system. This is a privileged situation, since it allows for a broad examination of alternatives. However, the person who analyzes and proposes recommendations must be knowledgeable, since the real situation does not yet exist and the analysis is based on hypothetical situations. These situations are predictions of real conditions, and today, thanks to advances in computer graphics, predicted scenarios can be simulated in the laboratory (Iida, 2005).

Ergonomics is often relegated to a post-design evaluation, leaving little opportunity for the ergonomist to make any significant and important design alterations. One way to change this scenario is to study the design process and, in particular, the ergonomics in designs (Burns and Vicente, 2000).

According to Santos et al. (2009), ergonomics was incorporated into the design process because of the need to consider human factors in designs. Rensink (2001) states that although the man-machine interface in industrial design is considered an integral part of engineering design, various misadjustments in operation and maintenance have occurred after the design has been implemented. This is because designers and builders have several points that must be met, including ergonomic factors.

For Van Wely (1970), the poor design of machinery, tools, furniture and other equipment, from the ergonomic standpoint, strongly affects the health of their users. To improve the ergonomic design of these elements requires greater integration between engineering and human sciences. According to Rensink (2001), designers are by nature more interested in the technical performance of their products. Moreover, because they do not operate or maintain facilities after the latter are put into operation, they do not have sufficient knowledge about the physical and mental environment that the user needs. Therefore, the application of ergonomic principles is easily forgotten during the design phases.

Haslegrave and Holmes (1994) argue that it is inevitable that differences between the designer's and ergonomist's concerns give rise to problems. The engineer or designer must concentrate on the structure and assembly of components, dealing with issues such as the availability, cost, strength and durability of materials. The ergonomist, on the other hand, must focus on the functions for which a product will be used and all the problems that users may have with it. Hence, the cross-purposes of these two different professionals may lead to conflicts.

Van Uden (2001) states that end users are often not consulted about the decisions taken during the design phases. Hendrick (2008), however, points out that ergonomics is not simply a matter of common sense among people, but requires the participation of qualified professionals.

Das and Sengupta (1996) presented systematic guidelines for the design of an industrial workstation. According to these authors, the implementation of their recommendations requires conformity between the anthropometry of the population and the various components of the workstation. Proper posture, work height, ideal work areas and visual requirements should be determined for the user population. Simpson and Mason (1983) concluded that ergonomic design produced by designers may be one of the most effective approaches to ensure the best application of the knowledge and practices of ergonomics.

Hendrick (2003) states that for industry and society, good ergonomics means good economy. According to Van Uden (2001), demonstrating the cost benefit of an ergonomics program usually motivates professionals to apply ergonomic principles. Auburn Engineers (cited in Adorno, 2004) argue that the cost of making adjustments to ergonomic requirements after a petrochemical plant has started its operations can represent 20% of the overall project cost, but that this cost represents only 0.5% if the ergonomic requirements are identified and solved in the design phase or during the definition of the design guidelines.

According to Iida (2005), ergonomics, like any other activity involved in the productive sector, is only accepted if it can be proven to be economically feasible. Therefore, the cost of applying ergonomic concepts is of paramount importance for the successful implementation of ergonomic design in large enterprises. Dul and Neumann (2009) conclude that significant changes must occur in the context of ergonomic research to shift from a health ergonomics paradigm to a business ergonomics paradigm, without losing sight of health and safety objectives.

Ben-Gal and Bukchin (2002) report that computerized tools are being used increasingly to design workstations. These tools offer two major advantages in the creation of designs: they enable the designer to examine a large number of design solutions and simulations, and allow the work task to be visualized in order to obtain various performance measures.

2. Object of Study

The object of study was a steam-raising boiler. The purpose of this kind of equipment is to transform liquid water into steam by absorbing heat produced by the combustion of a fuel (Pereira, 2002). A boiler is composed of two separate basic systems: the steam-water system, also called the water side of the boiler, and the fuel-air-gas combustion system, also called the fire side of the boiler (Gomes et al., 2003).

There are basically two types of boilers: fire-tube boilers and water-tube boilers. In fire-tube boilers, hot combustion gases pass through the tubes and the water from the boiler passes on the outside of the tubes, i.e., the fire side is on the inside and the water side is on the outside of the tubes (Gomes et al., 2003).

The main characteristic of water-tube boilers is that the steam is formed inside the tubes through which the water also circulates (Marques, 2005). According to the Brazilian standard NBR 11096 (ABNT 2000), water-tube boilers are characterized by the type of boiler in whose tubes water or steam circulates, with heat energy produced by combustion, which is applied to the external surface of the tubes.

Water-tube boilers are employed in a variety of large industrial units that use steam in their production process, such as the timber, chemical, petrochemical, textile, food, and power generation industries, among others. The metal and ceramic industries use steam to heat the fuel oil tubes that feed the furnace burners and to nebulize the fuel. The food industry uses steam for cooking food, while the textile industry uses large quantities of steam to heat its dyeing, bleaching and drying tanks (Gomes et al., 2003).

According to Vanelli (2004), chemical and petrochemical plants use steam for the following purposes:

- for driving the turbines of large machines such as compressors, blowers and turbogenerators;
- for driving pump and blower turbines;
- for heating pipes and product tanks;
- for heating products through heat exchange;
- for injection in towers for the adjustment of light products;
- as a driving fluid to generate vacuum in ejectors;
- for purging equipment to release and open them for maintenance;
- for cleaning equipment; and
- for decoking furnaces.

2.1. The Boiler Studied

In this chapter it was studied a water-tube boiler. The boiler was designed to produce steam at nominal conditions of 300 t/h x 91 kgf/cm²g x 485°C. If the boiler is operated above these limits it can cause serious damage and/or accidents.

The steam produced in this boiler is used in the industrial process to heat pipes and tanks, clean equipment, and drive equipment such as compressors and pumps, etc. In addition, the company's thermoelectric plant uses steam to produce electricity.

The boiler can be fired using the following fuels: refinery crude oil (REFCO), asphalt residue (AR), natural gas and liquefied petroleum gas (LPG). The fuel is transported through pipes to the burner, where it is mixed with combustion air, which causes it to ignite and produce the flame. Ignition is triggered automatically by the electric system, and the igniter is fed with natural gas or LPG. The furnace is surrounded by tubes that constitute the so-called "water wall." The combustion flame radiates heat, which is absorbed by the water wall.

Feedwater is introduced into the steam drum and passes down through the tubes at the rear of the tube bank and its side wall. These tubes distribute the water to the water drum, which in turn feed the tubes of the walls, floor, ceiling and front. The water and steam mixture reaches the steam drum through these tubes and the upper collectors of the furnace walls. The water and saturated steam circulate naturally within the tubes due to the difference in density between the mixture of water and steam resulting from the absorption of radiative (the surface of the furnace) and convective heat (at the surface of the furnace) on the boiler's heat exchanging walls. The saturated steam exits the steam drum and passes through the superheaters, after which it goes on to the process.

The final temperature of the combustion gases and their analysis are important factors to ensure ideal burning conditions and achieve optimal performance. These data are obtained by means of instruments installed at various points in the gas circuit.

2.2. Human Interventions in the Boiler System

2.2.1. Analysis of the Task

According to Gomes et al. (2003), the two most important variables to be controlled in water tube boilers are the steam pressure and water level in the tubes.

Satisfactory operation, availability of steam, and boiler operation time depend on the diligence and skill of the operator, who must be fully aware of his responsibilities, thus contributing to the smooth operation of the steam generating unit. The operator must also be fully qualified to intervene in case of an emergency.

Several adjustments must be made during the operation of the boiler, i.e., verification of the amount and distribution of air, fuel pressure, percentage of oxygen, and gas exit temperature, according to the load of the boiler. The smoke coming out of the chimney should also be checked to ensure that the quantity of air is adequate.

When starting up the boiler (at the beginning of the operation), it may be necessary to use all the burners to reach the desired output. However, the activated burners must be well distributed while the load is being increased to prevent uneven expansion. During normal operation, the load in the burners, as well as the pressure in the air boxes and the opening of air regulators must be uniform to ensure even distribution to the burners.

When the boiler is in operation, the feedwater should never be cut off, even briefly, because the consequences could be disastrous. Unless specified otherwise, the boiler's water drum level should always be maintained according to design specifications; this can be ensured by placing a mark of this working condition on the boiler's display screen. The water level should be controlled to keep it as stable as possible.

The soot blowers should be turned on periodically, as needed, to keep the heating surface clean. The soot blowers provide a rotating jet of steam inside the boiler, which removes the soot and deposits formed on the outer surface of the convection zone.

The elements of the superheater must be checked regularly for possible external excess fouling or any other melt, especially if the burned fuel has high ash content. Localized deposits can obstruct parts of the passage of gas through the superheater, providing the free regions a preferential channel for the passage of gases and thus raising the temperature of the elements of the superheater in this region. It is important for the external and internal surfaces of the superheater tubes to always be kept clean.

If the limits recommended for the conditions of the water are outside of normal specifications or if the treatment is inadequate, internal fouling may occur. Even in thin layers, this will cause augmented temperature in the tube wall and probable future breakage. Therefore, it is important to periodically inspect and control the water system to prevent this type of accident.

2.2.2. Description of the Activity

The boiler produces steam continuously and its operation is only shut down for maintenance or when there is little demand for steam for the process. The boiler operators work in 8-hour shifts, with five shift groups. There are always three groups working and two groups off duty.

The boiler is operated by means of interventions in the industrial area and through the control console. The area and console operators must communicate frequently to synchronize their activities. In this study, only the future work of the area operator was analyzed, since that of the control console operator was outside the scope of this work.

The boiler has several valves that are operated according to demand, comprising safety and relief valves, retention valves, instrument valves, drainage valves, and others. Some of them are controlled manually while others are motorized or automated. The manual valves are operated and have displays only in the industrial area, while the motorized valves are operated and have displays in the control console and the industrial area. The automated valves can be controlled and viewed via the console and in the area, and can be programmed to open or close automatically, as needed. If communication errors between the local system and the remote system occur, the motorized and automated valves must be operated manually by the area operator.

In addition to operating valves, the area operator must make visual inspections of the boiler to check for leaks, particularly in lines, verify the proper functioning of all the equipment and instruments, the flame in the burners (through inspection windows), check that the superheater is clean, etc.

Operations and situations found in inspections (equipment, leaks, valves, etc.), instrument readings and other information are recorded in the shift report. This detailed report is handed over to the operator taking over the next shift, giving him an overview of the status in the industrial area.

3. Analysis of the Boiler Design

Ergonomic Analysis of Future Work was used as the analytical tool. The application of this tool requires the participation of people knowledgeable about the boiler's operating processes. Furthermore, it should be noted that descriptions of future tasks often do not match the real activities that will be developed in industrial areas.

Thus, the formation of a multidisciplinary team of users and experts was crucial for the ergonomic analysis and the proposal of improvements. The users were operation technicians who will operate the boiler when its operation begins. The experts included technicians and engineers specializing in piping, civil engineering, automation, instrumentation and safety, as well as ergonomists and health professionals. This team was responsible for checking a variety of items in the design and assembly of the units, including the structural design, routing of pipes, installation of valves and lock valves, equipment, instruments, ergonomics and user safety and health. From the ergonomic standpoint, the team adapted the design to ergonomic concepts and oversaw the construction and installation to ensure their compliance with predefined standards and regulations. A computer model of the industrial unit was used as a graphical tool. This tool allows the design to be analyzed in three dimensions through graphic resources. Each team member had access to this tool at their own workstation. The implementation of the requirements was overseen and monitored according to the flowchart shown in Figure 1.

A tool was developed using Microsoft Office Access software to manage the database created by the multidisciplinary team. The fields that were created allowed for the inclusion of requirements according to title, date of creation, the person responsible for the requirement, location, specialty, classification, comments and images obtained through the computer model or on site. The requirements were monitored based on their status (concluded, ongoing), annotations (including the date and name of the person in charge) and images of the negotiated solution (in design) and their on-site implementation.

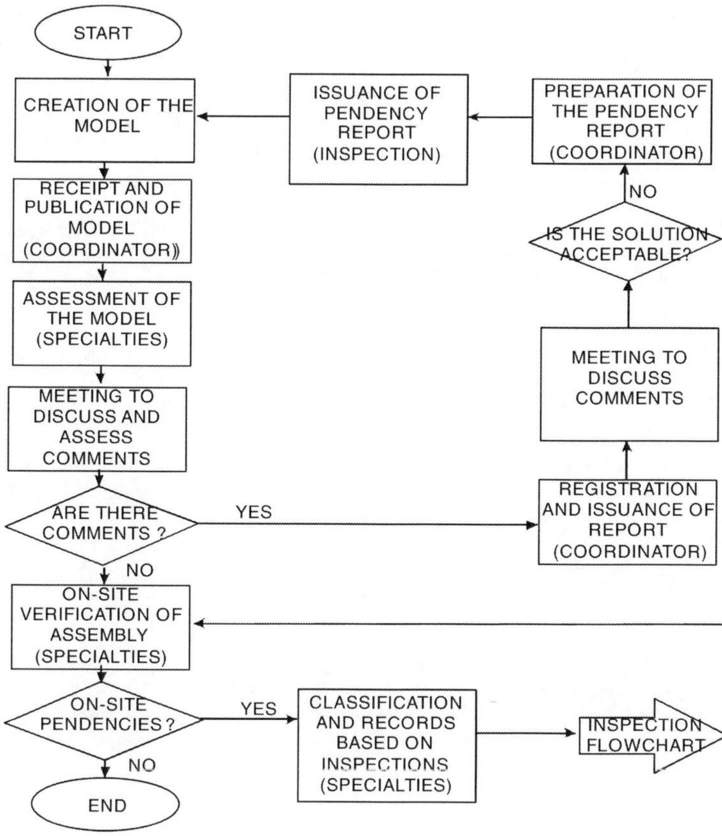

Figure 1. Flowchart of the analysis of the boiler design.

3.1. Ergonomic Requirements Found

To identify the ergonomic requirements, the multidisciplinary team held 19 meetings, with an average duration of 3 hours each, between March 2010 and August 2011.

During these meetings the data collected by each team member were compiled and discussed. The team identified 199 requirements resulting from nonconformities in various specialized areas. Among these requirements, 95 (48%) involved ergonomics.

Figure 2. Example of a requirement involving difficult access to the bypass valve.

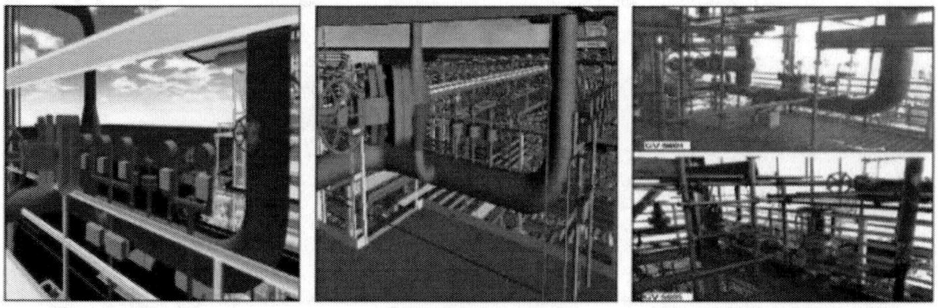

Image found Image corrected Photo of implementation

Figure 3. Example of a requirement involving difficult access to instrument valves.

Image found Image corrected Photo of implementation

Figure 4. Example of a requirement involving difficult circulation in the industrial area – excessively low platform ceiling.

Image found Image corrected Photo of implementation

Figure 5. Example of a requirement involving difficult circulation in the industrial area – non-compliant sailor's ladder.

The ergonomic requirements involved difficulty or impossibility in accessing valves, instruments, equipment, display screens and electrical panels (64 requirements – 68.5%) and difficulties in circulating in the industrial area (30 requirements – 31.5%).

The valves, instruments and equipment that were difficult or impossible to access included: the bypass valve (Figure 2), safety and relief valve, retention valve, instrument valve (Figure 3), drainage valves, junction box (JB), foundation fieldbus (FF), and others.

The main causes of difficulty in moving around (circulating) through the industrial area were: height (less than 2 meters) of the platform ceiling (Figure 4), the first rung of the sailor's ladder was irregular in relation to others and too high (Figure 5), the excessively narrow passageway and interference of the slanted ladder with irregular rungs.

CONCLUSION

The ergonomic study of a steam-raising boiler confirmed that the application of ergonomic design in a large industrial enterprise is possible and in fact essential. This study confirmed the importance of the participation of a multidisciplinary team composed of experts and future users in the assessment and proposal of solutions to ergonomic requirements, as well as the successful use of the analytical tool in designing for future needs. The flowchart for analyzing the design procedures and the software for managing the database proved to be suitable for monitoring of the ergonomic requirements identified by the team.

The requirements identified were found to involve physical ergonomics, which was visible in the analysis of the computer models; however, there was greater difficulty in identifying cognitive or organizational requirements in the detailing phase of the design of the industrial plant. All the recommendations for ergonomic improvements discussed and accepted by the multidisciplinary team were implemented on site before the start-up of the boiler's operation. The implementation of these improvements cost nothing to the company where the boiler was installed, since compliance with ergonomic principles was provided for in the contract.

REFERENCES

Adorno LR (2004). Levantamento de demandas ergonômicas em dez plantas petroquímicas brasileiras [*Survey of ergonomic requirements in ten Brazilian petrochemical plants*]. Porto Alegre, 170f. Master's Dissertation, Federal University of Rio Grande do Sul (in Portuguese).

Associação Brasileira de Normas Técnicas [*Brazilian Association of Technical Standards*] (ABNT, 2000). NBR 11096: Caldeiras estacionárias aquotubulares e flamotubulares a vapor – Terminologia. [*Stationary steam raising water-tube and fire-tube boilers – Terminology*] Rio de Janeiro, RJ, Brazil (in Portuguese).

Ben-Gal I, Bukchin J (2002). The ergonomic design of workstations using virtual manufacturing and response surface methodology. *IIE Transactions* 34(4) 375-391.

Burns CM, Vicente KJ (2000). A participant-observer study of ergonomics in engineering design: how constraints drive design process. *Applied Ergonomics* 31 73-82.

Das B, Sengupta AK (1996). Industrial Workstation design: A systematic ergonomics approach. *Applied Ergonomics* 27(3) 157-163.

Dul J, Neumann WP (2009). Ergonomics Contributions to Company Strategies. *Applied Ergonomics* 40(4) 745-752.

Gomes MD, Neto LZ, Almeida LRG (2003). SENAI – Serviço Nacional de Aprendizagem Industrial Escola SENAI "Prof. Dr. Euryclides de Jesus Zerbini" [*SENAI – National*

Service for Industrial Apprenticeship, SENAI School "Prof. Dr. Euryclides de Jesus Zerbini"] (in Portuguese).

Haslegrave CM, Holmes K (1994). Integrating ergonomics and engineering in the technical design process. *Applied Ergonomics* 25(4) 211-220.

Hendrick HW (2003). Determining the Cost–benefits of Ergonomics Projects and Factors that Lead to their Success. *Applied Ergonomics* 34 419–427.

Hendrick HW (2008). Applying Ergonomics to Systems: Some Documented "Lessons Learned". *Applied Ergonomics* 39 418–426.

Iida I (2005). Ergonomia: projeto e produção. [*Ergonomics: design and production*] São Paulo: Edgar Blucher, 2nd ed. (in Portuguese).

International Ergonomics Association (2000), Meeting of the Scientific Council, San Diego, CA, USA.

Marques FAP (2005). Modelagem e Controle de Nível do Tubulão de uma Caldeira de Vapor Aquatubular de uma Refinaria de Petróleo. [*Modeling and Control of the Level of the pipes of a water-tube boiler in an Oil Refinery*] Master's Dissertation. Postgraduate course in Engineering, Federal University of Rio de Janeiro – UFRJ, Rio de Janeiro, RJ, Brazil (in Portuguese).

Ministério do Trabalho e Emprego (1978). Portaria nº 3214. Norma Regulamentadora 17 (NR 17- Ergonomia) [*Ministry of Labour and Employment (1978), Ordinance No. 3214. Regulatory Standard 17 (NR 17 - Ergonomics)*] Brasília, DF, Brazil (in Portuguese).

Pereira CAA (2002). Curso de formação de operadores de refinaria: utilidades, sistema térmico e ar comprimido. [*Training course for refinery operators: utilities, heating system and compressed air*] Petrobras and UnicenP, Curitiba (in Portuguese).

Rensink HJT (2001). The Development of a Human Factors Engineering Strategy in Petrochemical Engineering and Projects: Part I. In: *Proceedings at the Occupational Health Offshore Conference*. Aberdeen 105-113.

Santos V, Zamberlan MC, Pavard B (2009). Confiabilidade Humana e Projeto Ergonômico de Centros de Controle de Processos de Alto Risco. [*Human Reliability and Ergonomic Design of Control Centers for High Risk Processes*] Rio de Janeiro: Synergia (in Portuguese).

Simpson GC, Mason S (1983). Design aids for designers: An effective role for ergonomics. *Applied Ergonomics* 14 (3) 177-183.

Van Uden MEJ (2001). The Development of a Human Factors Engineering Strategy in Petrochemical Engineering and Projects: Part II. In: Proceedings at the Occupational Health Offshore Conference. Aberdeen 115-129.

Van Wely P(1970). Design and disease. Applied Ergonomics 1(5) 262-269.

Vanelli CA (2004). Conservação de Água em Refinaria de Petróleo: O Caso da Revap. [*Water Conservation in an Oil Refinery: The Case of Revap*] Master's Dissertation. Postgraduate course, Federal University of Itajubá – UNIFEI (in Portuguese).

In: Noise and Ergonomics in the Workplace
Editor: Paulo Henrique Trombetta Zannin
ISBN: 978-1-62417-573-2

Chapter 7

ERGONOMICS AND HEALTHCARE WORKERS

Leila Maria Mansano Sarquis[1,*], Vanda Elisa Andres Felli[2] and Vivian Aline Mininel[2]
[1]Federal University of Paraná, Brazil
[2]University of São Paulo, Brazil

ABSTRACT

The World Health Organization reports that worker exposure to occupational risks leads to accidents, cancer, hearing loss, respiratory, musculoskeletal, cardiovascular, reproductive, neurological, dermatological, and mental disorders, which alone account for 1.7% of the years of life lost around the world. Countless consequences arise from healthcare workers' exposure to occupational risks. Work-related diseases resulting from risks in the work environment also occupy a prominent place. Musculoskeletal disorders represent the main consequence of exposure to ergonomic risks in the workplace. When subjected to adverse or inadequate conditions, the body responds negatively to the stimulus by manifesting pain, discomfort, illness and even death. This ergonomic study analyzes the extent to which the working conditions of healthcare workers are compromised, revealing that there are significant differences according to categories. Among these categories, nurses show a high prevalence of health problems, indicating that their working conditions are worse than those of other healthcare professionals. Several recommendations by international organizations that evince concern about the ailments of these workers must still be incorporated in different countries.

1. INTRODUCTION

Work is considered an organizer of society, of production and of the life of individuals. Under certain conditions, it promotes the lives of workers, enabling them to satisfy their consumer needs; however, under others, it facilitates exposure to certain risks that jeopardize

* E-mail: lmmsarquis@gmail.com.

their health, promoting their wear and tear. The working conditions that contribute to impair worker health include non-ergonomically friendly conditions.

Historically, ergonomics is conceived under different approaches, from the standpoint of studies based on the positivist paradigm.

The term ergonomics is of Greek origin and means labor laws: ergon = work and nomos = law. The term was first proposed by Polish scientist Wojciech Jastrzębowski in 1857, in his article "Tests of ergonomics, or the science of work, based on the objective laws of science over nature". The first definition established that "Ergonomics, as a science of work, requires that we understand human activity in terms of effort, thought, relationship and dedication" (Jastrzębowski cited in Karwowsky, 1991).

Vidal (2011) states that Jastrzębowski's concept of ergonomics, at the time, emphasized the mobilization of: "... four aspects of psychic nature, namely, the physical-motor nature, sensory-aesthetic nature, the mental-intellectual nature, and the spiritual-moral nature.

This science of work therefore signified the science of effort, play, thought and devotion. One of Jastrzębowski's basic ideas is the key proposition that these human attributes deflate and decline due to their excessive or insufficient use".

However, it was in the post WWII movement that ergonomics began to be conceived as we understand it today, as a result of the foundation of the first society of Ergonomics, in 1947, the Ergonomics Research Society. It was also at this time that the stream of ergonomics called Human Factors Engineering, or HFE, emerged (Vidal, 2011).

2. The Conception of Ergonomics and Its Approaches

In 2000, the International Ergonomics Association (IEA) adopted the following official definition: "*...it is a scientific discipline related to understanding the interactions between humans and other elements or systems, and to the application of theories*, principles, *data and methods in designs in order to optimize human well-being and* the *overall performance of the system. Ergonomists contribute to the planning, design and evaluation of tasks, workstations, products, environments and systems to make them compatible with people's needs, abilities and limitations*" (Brazilian Ergonomics Association, 2012).

Thus, the Brazilian Ergonomics Association – ABERGO, considers ergonomics (or human factors) a discipline oriented toward a systemic approach to all aspects of human activity. To encompass the magnitude of this dimension and enable them to intervene in work activities, ergonomists must adopt a holistic approach to the entire field of action of the discipline, in terms of its physical and cognitive aspects, as well as its social, organizational, environmental and other aspects (ABERGO, 2012). ABERGO considers that, in general, the domains of expertise of ergonomics are:

- *Physical ergonomics* – relates to the characteristics of the human anatomy, anthropometry, physiology and biomechanics in their relation to physical activity. Relevant topics include the study of working postures, handling of materials, repetitive movements, work-related musculoskeletal disorders, workstation design, safety and health.

- *Cognitive ergonomics* – refers to mental processes such as perception, memory, reasoning and motor response as they affect the interactions between humans and other elements of a system. Relevant topics include the study of mental workload, decision making, specialized performance, human computer interaction, stress and training as they relate to designs involving humans and systems.
- *Organizational ergonomics* – concerns the optimization of sociotechnical systems, including their organizational structures, policies and processes. Relevant topics include communications, crew resource management (CRM – aeronautical), work design, temporal organization of work, group work, participatory design, new work paradigms, cooperative work, organizational culture, networking organizations, telework, and quality management.

Almeida (2011) contends that the application of ergonomics dates back to prehistoric times, when the caveman began to adapt the environment, his weapons and his tools to his anatomical characteristics. He also states that ergonomics became the object of study from the moment when humans felt the need to adapt to new work schemes generated by the increasing changes in production processes. This led to the development of two approaches to ergonomics: the Anglo-Saxon or classical approach – led by the Americans and British, and the French or contemporary approach – practiced initially by the francophone countries, which subsequently became universal. These approaches were based on the different conceptions of English and French measurement systems.

Anglo-Saxon or classical ergonomics emerged as a structured discipline soon after World War II, at which time the following definition of ergonomics was adopted (Almeida, 2011): "*The study of man's relationship with his work, equipment and environment, and particularly the application of knowledge about anatomy, physiology, and psychology in solving problems arising from this relationship*".

This ergonomics is based on the mechanical/thermodynamic paradigm of humans (Almeida, 2011). According to Montmollin (1990), this approach considers Ergonomics as the use of certain sciences to improve the conditions of human labor and focuses on methods and technologies, since what matters are the physical aspects and human capacities in the man-machine relationship, i.e., the need to adapt the machine to man. This approach focused on aspects such as force, posture, repetition, and range, using knowledge from different areas such as anthropometry, psychology, physiology and biomechanics (Almeida, 2011).

Another approach is the French or contemporary ergonomics, which appeared more recently in the mid-20th century in France. The purpose of this approach (Almeida, 2011) is to adapt work to man, focusing on the specific study of human labor, which requires an analysis of the activity performed, directing the study on the interrelationship between man and the production environment in which he is inserted. This approach gave rise to ergonomic work analysis (EWA).

Montmollin (1990) states that EWA is a stage in ergonomic intervention, which involves addressing the task (what to do – prescribed work) and the activity (what is done – real work), demonstrating the core concern with work organization. EWA answers the following questions: What is done? Who does it? How is it done? How could it be done?

In short, classical ergonomics focuses on the human factor, considering Man as a component of the work system, while contemporary ergonomics focuses on human activity, considering Man as an actor of the work system. In this sense, the two approaches are complementary in practice.

Regardless of the school of thought, there is a worldwide recognition that workers are subject to occupational health hazards arising from work performed in inadequate conditions.

The World Health Organization (WHO, 2009) reports that worker exposure to numerous occupational hazards leads to accidents, cancer, hearing loss, and respiratory, musculoskeletal, cardiovascular, reproductive, neurological, dermatological, and mental disorders, which, alone are responsible for 1.7% of the years of life lost (DALYs = *Disability Adjusted Life Years* (DALYs): The sum of years of potential life lost due to premature mortality and the years of productive life lost due to disability) around the world.

Despite the difficulty of obtaining data, WHO also reports that over 350,000 workers lose their lives each year due to work-related health problems, with major differences between countries with high and low per capita income. With regard to ergonomic aspects, the report states that 37% of back pains are attributed to work, such as handling heavy weights, physical demands of work, frequent bending, twisting and awkward postures. The report also points out that, albeit not a cause of mortality, low back pain causes considerable morbidity and is the greatest cause of absenteeism, resulting in economic loss. Punnett et al. (2005) estimate that work-related back pain causes an annual loss of 818,000 years of productive life due to disability.

Concha-Barrientos et al. (2005) reported that work-related diseases caused 775,000 deaths worldwide in 2000, and that the occurrence among men was five times more frequent than among women. The most frequent cause was accidents (41%). The authors also reported that workers who became ill due to exposure to occupational hazards lost 22 million years of healthy life. On the other hand, Nelson et al. (2005) reported that, in the same year, in addition to back pain (37%), worldwide worker exposure to occupational hazards was responsible for hearing loss (16%), chronic obstructive pulmonary disease (13%), asthma (11%), accidents (8%), lung cancer (9%) and leukemia (2%). They calculated that these risks accounted for 850,000 deaths and 24 million years of healthy life. The same authors also reported that sharps injuries accounted for 40% of hepatitis B and C infections and 4.4% of HIV infections in healthcare workers.

Worker exposure to occupational hazards, especially those related to ergonomics, represent a major impact on the global economy, which could be substantially reduced through prevention. Scholars and international organizations (Nelson et al., 2005; Cole et. al., 2006; WHO, 2009; Occupational Safety and Health Administration - OSHA, 2012) have attributed the reduction of the economic impact of diseases to the implementation of ergonomic programs in the workplace and to other preventive strategies associated with the reduction of frequent and severe pain of the workforce.

In this regard, the exposure of healthcare workers to occupational hazards, which are of interest to ergonomics, can be evidenced by biological, physical, chemical, ergonomic and psychosocial risks (OSHA, 2012a). The Occupational Safety and Health Administration (OSHA, 2012a) presents the following definitions:

- *Biological risks*: "Apart from the risks of caring for patients with acute infectious diseases, the most important biological risks are viruses causing hepatitis and AIDS (acquired immunodeficiency syndrome). The latter diseases are associated with invasive ('bloody') interventions." (OSHA, 2012a).

- *Physical risks*: "The most significant physical risk in health care is ionizing radiation in the form of X-rays and radioisotopes. Risk from these sources has been reduced considerably because of strict regulations, engineering of modern technology and the use of personal protective equipment." (OSHA, 2012a).
- *Chemical risks*: "The most dangerous chemical risk factors in health care are drugs used in the treatment of cancer. Most of these so-called chemotherapy drugs are able to induce cancer or damage reproduction, thus there is no safe value for them. Workers are exposed to these substances through skin contact or inhalation. The most high-risk activity in this regard is the preparation of chemotherapy drug solutions, a process that must be carried out in isolated conditions. Administration of these solutions and work with patient excreta and contaminated linen also calls for adherence to safety instructions such as the use of gloves and aprons." [...] In addition, "...disinfectants and cleaning substances that can cause skin diseases. Gloves contain latex, which can be a source of allergy. Laser, ultraviolet light and electromagnetic fields are far less common physical risks. Physical work affecting the muscles, joints and bones, and especially the lower back, is associated mainly with patient handling. Nurses and paramedics are at the highest risk of developing musculoskeletal disorders (MSDs). Adverse ambient conditions can pose a risk in the case of heavy work and high temperatures, e.g. in operating theatres" (OSHA, 2012a).
- *Psychosocial risks*: "The unpredictability of events, everyday tragedies, night shifts, the stressful work of an emergency unit, aggressive behavior by patients, violence from frustrated relatives and working abroad are the most common psychosocial risk factors. These circumstances can place a heavy burden on health care workers and give rise to psychological problems (e.g., stress, burnout) or psychosomatic disorders. Among those particularly vulnerable to mental stress are care workers in wards with few successful outcomes (e.g., social care for the severely disabled, oncology ward) or where tragedies happen frequently (e.g., emergency department." (OSHA, 2012a).

3. Contextualization of Health Services

The production of health services is included in the service sector of the economy of capitalist countries. As such, healthcare is organized in a specific way and differentiated from other types of work.

Healthcare aimed at controlling diseases on a social scale and at recovering the disabled workforce (Gonçalves, 1994) is cooperative work performed in conjunction among different types of work, using different areas of knowledge and practices, which characterizes a social and technical division of worker teams. Healthcare workers are nurses, doctors, social assistants, psychologists, physiotherapists, occupational therapists and others who deal with people in situations that involve demands in their health-disease process. Dealing with suffering, death and pain is a specificity of these workers and has caused not only suffering and pleasure but also stress, burnout, anxiety, depression and other health problems. These workers are part of the salaried or self-employed labor market. As salaried workers, they are employed in hospitals, in public health systems, etc., and as self-employed workers, they have private clinics, offices, etc.

Healthcare workers employed in health institutions are organized in teams; work in shifts, including night shifts; often handle heavy weights; adopt awkward postures; use furniture and high-tech equipment which is also inadequate; work long hours under stress; travel long distances and in fast-paced activities; and in other forms of rigid work organization, which are insalubrious, dangerous and, in many countries, also precarious. This scenario represents a high workload which, combined with an inadequate environment, diminishes the significance of the work content, leading to dissatisfaction, demotivation and other irksome processes for workers (Carayon, 2012).

These situations experienced by healthcare workers require physical, cognitive and organizational ergonomic interventions, implying the need for improved work, life and health conditions for the worker. For companies, good ergonomic work conditions translate into higher productivity and lower work-related health costs.

4. Occupational Exposure of Healthcare Workers and Ergonomic Risk in Different Forms of Work

The inadequate working conditions and exposure to ergonomic risks to which nursing workers are subjected on an international level have led to frequent reports of musculoskeletal symptoms and WRMSDs (Vilar, 2002; Célia and Alexandre, 2004; Varela, 2004, Murofuse and Marziale, 2005; Larese and Fiorito, 2007; Leite et al., 2007; Loiola and Felli, 2007; Barbosa, 2008). These disorders are broadly addressed in studies, as well as disorders of the neck and shoulders (Smedley et al., 2003; Trinkoff et al., 2003), lower back (Smedley et al., 1995; Roupa et al., 2008); elbows (Palmer et al., 2000), hands and wrists (Surawera, 2011); hips (Eriksen, 2003); and knees and feet (Tinubu et al., 2010; Oliver, 2009).

The study entitled "Psychosocial and Cultural Influences on Disability – CUPID" (Felli et al., 2011), which involved 192 nursing workers, revealed that about 63% of these workers reported musculoskeletal symptoms in the month preceding the study, while 74.6% reported low back pain, 76.5% reported neck pains, 80.2% complained of pain in the shoulders, 86.1% in the elbows, 84.6% in the wrists/hands and 78.2% in the knees. Brazilian and Italian nursing workers (Bonzini et al., 2011) present a similar prevalence of pain in various parts of the body, although Italian nurses present a higher prevalence of symptoms in the neck (32%) and higher absenteeism due to lower back pain (21%) when compared to Brazilian nurses.

WRMSDs are associated with psychosocial and gender risks, since comorbidities between musculoskeletal disorders and depression, stress and other mental disorders have been reported in the literature (Magnago et al., 2009), as well as a higher incidence of musculoskeletal disorders among women (Wijnhoven et al, 2006; Nicot, 2008; Gjesdal et al., 2011).

Other risks arising from shift work subject nursing staff to diseases such as cancer, diabetes, cardiovascular disease, Parkinson's disease, stomach problems, insomnia, tuberculosis and other disorders (Chen et al., 2006; Souza and Bertolozzi, 2007; Abdalkader and Hayajneh, 2008; Pan et al., 2011; Shouldice, 2011, Karino, 2011).

Studies in this field focusing on pharmacists are aimed at verifying possibilities of risk prevention and health promotion.

The work performed by this health professional involves the direct manipulation of pharmaceutical formulations, monitoring of infectious diseases, analysis and epidemiological research of laboratory tests, sanitary and medication surveillance, and the task of storing, transporting and distributing pharmaceutical products such as foods, cosmetics, immunobiological and correlated products. The pharmacist's work also involves handling laboratory balances, fume hoods, distillers, incubators, microscopes, glassware, etc., and many of these activities involve exposure to biological, chemical and ergonomic risks. Therefore, concern should focus on safety factors in terms of the size of the workplace, the placement of objects to ensure safe handling and manipulation, a well ventilated environment, a bench height that ensures that ensures minimal effort by the rampant worker, guidelines for the proper handling of equipment, rest pauses, improvements in the organization of activities aimed at productivity relating physical to mental comfort, generating efficiency and avoiding such exposures (Ministry of Labor and Employment, MTE, 2010). Research into this professional category found that pharmacists reported work-related musculoskeletal disorders. After confirming the exposure to ergonomic risk, measures were adopted to implement ergonomic interventions in the productive process, which diminished the discomfort and functional overload. However, other measures must be implemented to reduce the repetitiveness and adjust the scale of the production process in order to improve ergonomic conditions at work (Borin, 2004; Nyssen and Côte, 2010). Such measures should be implemented and incorporated in the work process, minimizing such injuries to the worker's health.

The emerging risk among social workers is linked to their activities of guidance to individuals, families and communities regarding duties, standards, codes, and legislation. These workers plan and organize social programs, provide guidance for and develop activities aimed at developing citizenship. They may work in public or private institutions and Non-Governmental Organizations (NGOs), and their work involves teams, indoor environments, and daytime work hours, although it may occasionally involve working irregular hours during shifts and emergency cases, in stressful situations and under pressure, which may trigger exposure to various occupational hazards, including ergonomic ones. Studies conducted in Italy with social workers show the existence of high prevalence rates of musculoskeletal disorders among these workers, particular symptoms of low back pain and cervical pain. Another relevant aspect revealed by this study were complaints resulting from psychosocial pressures, such as depression and anxiety. These findings indicate that musculoskeletal disorders are related to stress factors of the process of work organization and also to individual emotional factors. The preventive measures proposed in this study to prevent work-related disorders in social workers were educational, organizational and ergonomic (Mastrangelo et al., 2008; Magnavita, 2009).

Brazil's CNAE describes medical doctors as health professionals who provide medical consultations and treatment of patients, implementing actions aimed at preventing disease and promoting health, individually and collectively. Medical doctors may work in sectors where their actions benefit collective activities of health education and social services; they may act autonomously; their activities are performed indoors; they may have highly irregular working hours; perform their work in awkward and uncomfortable positions for long hours, and they are subject to constant stress. In some cases, physicians may handle toxic, chemical, biological, and radioactive materials, and be exposed to other hazards, including ergonomic risk (Ministry of Labor and Employment, MTE, 2010). Research among physicians has found the presence of ergonomic risk in the work process; however, proper interventions minimize such risk exposure (Boiselle et al., 2008; Wahlstedt et al., 2010; Healy et al., 2011).

A recent study involving medical doctors conducted by the Pan American Health Organization (PAHO, 2012) in Argentina, Brazil, Costa Rica and Peru found that these professionals are subject to ergonomic risk exposure due to work overload, working conditions and hours worked. The study found that the majority of physicians (57%) work in more than one workplace, and in Costa Rica they work on average more than 48 hours a week, which further compromises the life of these professionals (PAHO, 2012).

Veterinary doctors engage in the treatment of animal health in veterinary clinics, develop research projects, and publish results and information. Veterinarians may work in public or private companies and laboratories in activities involving agriculture and animal health (Ministry of Labor and Employment, 2010). The professional activities of veterinarians tend to expose them to pathogens and toxic, chemical and biological materials as well as to ergonomic risk (Labarthe and Pereira, 2008; Lucas et al., 2009).

Research into the health of dental professionals (dentists and dental surgeons) is aimed at evaluating the risks in this field of activity and the possible impairments and diseases associated with this exposure.

These professionals are exposed to physical hazards, particularly to noise outside the workplace and to the noise generated by their own dental equipment, e.g., high and low speed rotating motors, compressors, suction machines and others (Hinze et al., 1999; Nogueira et al., 2010). Under normal working conditions, noise levels range from 60 to 70 decibels (dB), but the sensation of discomfort increases between 70 and 90 dB, and noise levels above 90 dB lead to the risk of hearing loss (Nogueira et al., 2010). Under Brazilian labor laws, Regulatory Standard 15, which governs insalubrious activities and operations, specifies 85 dB as the maximum tolerable noise level during an 8-hour workday (Ministry of Labor and Employment, MTE, 2001b).

The literature also highlights exposure to chemical risks, especially to medicinal gases (such as nitrous oxide) and mercury (associated with dental amalgam fillings), the latter being more severe due to its toxic potential, cumulative effect in the body, and difficulty to manage and control (Grigoletto et al., 2008; Nogueira et al., 2010). In 2008 and 2009, Norway, Sweden and Denmark banned the use of mercury amalgam, and in the same period, some American states also passed laws requiring the informed consent of patients who receive dental restorations (Directorate for Health and Social Affairs of Norway, 2003; Edlich et al., 2009). Currently, the website of the Food and Drug Administration – FDA contains information about the potential neurotoxicity of mercury in dental amalgam, stating that it may impair fetal and infant development and that it is mandatory for dentists and patients to discuss its use (Edlich et al., 2009; Food and Drug Administration, 2012).

Biological risks associated with contact with the oral cavity, which is rich in microorganisms, increase the possibility of contagion of diseases such as hepatitis and AIDS. Ergonomic risks result from incorrect postures, excessive work pace, repetitive actions, and other factors. Mechanical risks, in turn, are characterized predominantly by sharps injuries caused by handling surgical and dental instruments. Lastly, social risks pertain to the stresses generated in the relationship between the professional and the patient (Souza, 1997; Garbin et al., 2009; Lopes et al., 2012).

Several diseases are identified as common in this occupational group, such as kyphoscoliosis; repetitive strain injuries (RSI) and work-related musculoskeletal disorders (WRMSDs), especially in the upper limbs, shoulders, spine, neck and cervical spine; noise-induced hearing loss; contamination by mercury and by ionizing and nonionizing radiation; allergies; and contact dermatitis and eczema (Costa et al., 2006; Nogueira et al., 2010).

Although their work focuses on the recovery and rehabilitation of patients with musculoskeletal problems, physical therapists also belong to a professional category that shows a high prevalence of WRMSDs. Because their body is their work tool, they are exposed predominantly to ergonomic risks associated with the manipulation of patients and with activities that require the application of force, repetitive movements and awkward postures (Campo et al., 2008; Passier and McPhail, 2011).

The high prevalence of musculoskeletal disorders related to the working conditions of young physical therapists has been addressed globally, revealing a high incidence of lower back pain (Cromie et al., 2000; Nyland and Grimmer, 2003; Obembe et al., 2008). Several studies have found a high rate of musculoskeletal disorders among physical therapists in different regions of the world: 78% in Nigeria, Africa (Obembe et al., 2008), 71.6% in Kuala Lumpur, Malaysia (Nordin et al., 2011); 85% in Izmir, Turkey (Salik and Özcan, 2004), 91% in Australia (Cromie et al., 2000), 49.2% in Edmonton, Canada (Mierzejewski and Kumar, 1997), 32% in the United States (Holder et al., 1999); and 90.7% and 71% in regions of Rio Grande do Sul and Minas Gerais, Brazil (Souza D'Avila et al., 2005; Pivetta et al., 2005).

Among the most affected areas are the cervical, dorsal and lumbar spine, shoulders, hands/wrists, the thoracic area, and feet/ankles, with women being more susceptible to injury (Pivetta et al., 2005; Leite et al., 2010). Moreover, the most common injuries in this professional category have been found to be tendinitis, ligament sprain and muscle strain, intervertebral disc degeneration, sprains, degenerative disorders, tendinosis and synovitis (Romani, 2001).

Despite the high prevalence of WRMSDs among physical therapists, few professionals miss a day's work or take a leave of absence due to pain, which may be attributed to several factors, i.e., their commitment to patients, job instability, or self-treatment performed as a strategy for pain management (Souza D'Avila et al., 2005; Bork et al., 1996).

Some studies have proposed strategies for the prevention of pain and musculoskeletal injuries, particularly those related to manual therapies, such as the correction of body mechanics, adjustment of work equipment, frequent postural changes, suitable exercise, and even changes in professional practice (McMahon et al., 2006; Passier and McPhail, 2011; Sharan and Ajeesh, 2012;).

The health of workers in the field of psychology has also been the subject of several studies, mainly as it pertains to their exposure to psychic risks and burdens, which are characteristic of this professional occupation. The possibility of identification with patients and the formation of emotional bonds between psychologists and their patients (Covolan, 1996); maintenance of the therapeutic relationship; excessive involvement in work and personal burnout (Faber, 1985), as well as loneliness, expectations and lack of satisfaction (Rabin et al., 1999) are predisposing factors of occupational stress.

Concerns about specific patients, the emotional impact caused by severe cases, and problematic relationships with coworkers were cited as the most prevalent internal stressors in a population of psychologists in the state of Paraná, Brazil. The external stressors cited by these psychologists were long working hours, financial instability and lack of breaks during the workday (Sanzono and Coelho, 2007).

A survey of psychologists of the Portuguese Association of Psychologists found that women suffer higher levels of stress, burnout and professional dissatisfaction than male psychologists (Gomes and Cruz, 2004).

Several studies to analyze stress and burnout among psychologists have found correlations between the many variables present in the daily work routine (Benevides-Pereira, 2002; Benevides-Pereira and Moreno-Jiménez, 2003; Biehl, 2009). All the findings point to a high prevalence of these conditions among the analyzed professionals, which indicates the need for intervention in this occupational field.

Psychologists use several coping strategies to deal with conflicts, stress and overload in their workday routines, such as periods of rest and vacations; concentrating on a patient's problems only during the session, and then "disconnecting" from the clinic when at home; reduction of working hours; establishment of clear agreements about the functioning of therapy; stress management; seeking help from third parties or through entertainment, among other strategies (Covolan, 1996; Sanzono and Coelho, 2007).

No studies were found in the literature about the risks and workloads of physical educators and personal trainers, or about the health and disease processes of these workers. Possibly, since it is a profession associated with regular physical exercise and healthy eating habits, this professional occupation has not yet been the focus of studies. However, it should be noted that the practice of excessive exercise may lead to damage and injuries of the musculoskeletal system, culminating in a process of intense physical fatigue or burnout. Moreover, this field of activity has its own peculiarities and involves different relationships, thus deserving a closer scrutiny.

5. Consequences of Occupational Exposure on Health

Healthcare workers suffer countless consequences as a result of their exposure to occupational hazards. Work accidents are the most easily identified, since they affect the worker's physical body, leaving behind visible and sometimes irreversible marks. Work-related diseases and injuries also stand out as a result of risks in the work environment.

Musculoskeletal disorders represent the main consequence of exposure to ergonomic risks in the workplace. When subjected to adverse or inadequate conditions, the mechanics of the body responds negatively to the stimulus, which is manifested by pain, discomfort, illness and death.

The tendency for chronicity to work-related musculoskeletal disorders (WRMSDs), besides constant pain and discomfort, generates a feeling of impotence and of diminished quality of life and of social relationships in the worker. However, the most important consequence of these disorders is temporary or permanent disability, which may lead to early retirement.

Currently, the most widely used definition of the term disability comes from the International Classification of Functioning (ICF) recommended by the WHO, which standardizes the language for health, disability and functionality-related issues (WHO, 2002).

According to this concept, the term disability acquires a global meaning that encompasses not only disability (body functions and structures) but also limitations for activities and constraints in participation (WHO, 2002). This concept tends to integrate the medical and social models of the approach to disability, which is defined by the social model as a biopsychosocial synthesis that offers an integrated and coherent approach to the different dimensions of health, from a biological, individual and social perspective (Rossi et al., 2007).

Despite the various laws on work-related disability around the world – worker rights, benefits, models of professional rehabilitation and return to work – the consensus remains that the temporarily disabled worker must be rehabilitated for work, and in permanent cases, should be retired from it.

In general, people with musculoskeletal disorders present substantial levels of disability, poor rates of return to work and high socioeconomic costs associated with the time off from work. The psychosocial variables also clearly influence the rehabilitation process of injured workers and return to work outcomes (Lloyd et. al., 2008).

The international literature discusses the social consequences of occupational injuries and illnesses, highlighting the losses to workers who suffer them, which involve not only pain but also reduced income, loss of productivity, unemployment, stress, depression, isolation, stigma and discrimination, impairment of family and sexual relationships and the adoption of risk behaviors such as alcohol and drug abuse (Dembe, 2001; Keller, 2001).

This situation reaches greater proportions when viewed in the light of gender, where women (who predominate in some healthcare professions such as nursing) are more affected, including the development of their household activities and the care and education of their children.

Absenteeism and presenteeism are topics increasingly associated with the consequences of work-related musculoskeletal disorders (WRMSDs), which affect all the stakeholders involved in the workplace: workers, coworkers, clients and employers. For the workplace, the consequences of absenteeism are as relevant as those of presenteeism, i.e., work overload for workers who are present at work or whose performance is preserved. For the employer, absenteeism means the loss of effective and active manpower, and in Brazil, if absenteeism does not exceed 15 consecutive days, it means paying in full a worker who was not fully at work.

Absenteeism is characterized by unplanned absences from work, arising from health-related absences and sick leave; this concept excludes days off, vacations and other planned and predictable leaves of absence (Gaidzinski, 1991). Presenteeism, in turn, is related to job performance and productivity. By definition, it means being present at the workplace, but unable to produce fully due to illness or other disabling conditions (Hemp, 2004). For this author, presenteeism may represent a reduction of individual productivity by up to one third or more.

Both situations cause losses in the context of work and worker health. In the case of absenteeism, many workers do overtime to try to compensate for absences due to illness (Kessler et al., 2004), which will prolong the baseline situation that triggered the initial absenteeism, deeply affecting their health and relationships with other workers.

For the work context, absenteeism represents an overload for coworkers and impairment of the work organization. In the case of health professionals, who frequently work in health institutions and in teams, this represents the fragmentation of activities and shortcomings in the assistance provided to the patient. For the employer, it means augmented indirect costs resulting from loss of productivity due to absenteeism or early mortality (Nita et al., 2010).

In Brazil, while he receives paid sick leave due to occupational injuries, the worker is considered to be on leave and is entitled to 12 months of job stability after returning to work (Ministry of Labor and Employment, MTE, 1991, 2001a). However, this partial stability does not prevent the worker from being dismissed at the end of the 12-month term, a fact that often causes him to return to work before the end of established period of sick leave, in an attempt to avoid dismissal.

In the case of presenteeism, the worker continues to work (does not absent himself or take sick leave), but without being fully able to perform his tasks. Thus, since his health problem is of occupational origin, remaining in this environment means persisting in his exposure to occupational risk and the potential worsening of the case.

Unlike absenteeism, presenteeism compromises worker performance and productivity, putting at risk the assistance provided to ensure the health of clients due to the likelihood of failures and omissions. A loss of productivity, even in low percentages, is a sign of the subtle coexistence of interferences and consequences, be it to the organization of the service and the institution or to the professionals and the assistance rendered (Umann et. al., 2012).

In an attempt to continue working, even when not in good health, the health professional uses strategies to minimize his pain, which include the consumption of large quantities of drugs such as analgesics to relieve the symptoms. Under these circumstances, absence from work or submitting to medical care only occurs in extreme situations, when the health impairment has worsened, and in some cases, is irreversible (Mininel, 2010; Loiola and Felli, 2007).

Within this context there are also the conflicts established between workers and health insurance companies – for the granting or renewal of benefits, and with the employer – in most cases, to prove the causal link between the disease and the occupational exposure. This scenario tends to further contribute to the suffering of the worker, especially from the psychosocial standpoint.

Work is a form of social identification and the loss of this identity, allied to the loss of capacity, is a very painful process for the individual. The analysis of the individual in his work context is necessary for the proposal of actions directed at promoting health and quality of life. Work cannot be dissociated from life, and therefore, this should be considered a determining factor for certain conditions of health and disease.

6. Prevention of Exposure to Ergonomic Risk

The exposure of healthcare worker to ergonomic risk and its effects in terms of both the health problems it creates and their consequences implies the need to establish preventive measures to ensure the workers' right to health and productivity in the effectiveness of the healthcare afforded to individuals.

As stated in the Seoul Declaration on Safety and Health at Work, the right to a safe and salubrious workplace should be recognized as a fundamental human right, reiterating the principle whereby this guarantee is understood not only with respect to existing risks but also to emergent risks. Although it is often difficult to grasp the correlation between work and health, it is tacitly understood that they are interdependent, since performing work requires being healthy and in many cases work is the cause of illnesses, while in others it leads to satisfaction and personal fulfillment (International Labor Organization -ILO, 2010).

As a result of recommendations from international agencies concerning the health of healthcare workers, guidelines specific to each country have been proposed, as in the case of Brazil, which are described below.

Brazil's National Occupational Health and Safety Policy (PNSST) establishes guidelines for the improvement of working conditions and the prevention of worker exposure to physical and mental impairment. The PNSST aims to reduce work-related injuries, i.e., their prevention at different levels through promotion, rehabilitation and surveillance. Its guidelines include comprehensive healthcare, and interlinked intra- and intersectoral action, worker participation, the support of studies and the training of human resources. This interlinked action involves the Ministries of Labor, Social Security, and Health, and enables the reporting of incidents between formal and informal work, as occurs in this country, taking advantage of the infrastructure of the Unified Health System (SUS) (Brazil's National Health Service), the National Network for Workers' Inclusive Healthcare (RENAST) and the Information System for Notifiable Diseases (SINAN) (Ministry of Labor and Employment, MTE, 2007).

With regard to the promotion of health as it pertains to ergonomic risk, Brazil's Ministry of Labor has established Regulatory Standards, which serve as legal measures for the prevention and correction of existing and new hazards, providing guidelines for mandatory procedures related to safety and occupational medicine. The role of these standards is to ensure that work is developed in a healthy and effective manner and to enable the employer to foresee risks and to adapt them as much as possible to the worker's activities. Regulatory Standard 17, which pertains to ergonomics and work organization, establishes guidelines about their suitability to the psychophysiological characteristics of workers and to the nature of the work to be performed. For the purposes of this regulatory standard, the production standards of work achievement, time and pace should be taken into account (Ministry of Labor and Employment, MTE, 2002).

Other regulatory standards prescribe rules for periodic medical examinations, the requirement for internal accident prevention committees, the use of individual and collective protection equipment, the evaluation and monitoring of safe electrical installations, risk management, and other legal requirements complemented over the years, currently making a total of 35 regulatory standards (Ministry of Labor and Employment, MTE, 2001b).

With regard to healthcare workers, Regulatory Standard 32, created in 2005, establishes basic guidelines for the implementation of health and safety protection measures for health service workers, as well as those engaged in health promotion and assistance activities in general, and requires employers to provide Individual Protection Equipment (IPE) commensurate to the risk the worker is exposed to. The efficiency of this equipment must be such that it ensures control of the exposure risk as well as the comfort of the professional. This equipment, whether or not disposable, should be readily available in sufficient numbers to ensure immediate delivery and/or replacement. Employers must also provide training in the use of IPE for their employees, who in turn must use IPE correctly and at all times (Ministry of Labor and Employment, MTE, 2005a). The law provides for exposure to various risk factors that are present in the workplace, such as biological, chemical and physical risk factors, with emphasis on ionizing radiation, as well as on ergonomic risk factors. With regard to ergonomics, the standard addresses the need for maintenance of the equipment and mechanical devices used for transport and at the beds of patients; proper conditions of comfort in terms of noise, lighting, thermal comfort, cleanliness and conservation; the provision of safe devices in workstations, whose stability enables workers to reach high objects without requiring extra effort; the use of devices that minimize effort in transporting and moving equipment and patients, aided by mechanical or electromechanical means. In addition, this regulatory standard establishes the proper body mechanics when moving patients or equipment; measures to be taken when dealing with patients with behavioral disorders; and the use of odor exhaust devices.

In Brazil, up until 2007, it was the worker's responsibility to provide proof of risk exposure and of resulting work accidents and/or disorders. However, the creation of the Technical Epidemiological Nexus (NTE) in 2007 changed the legal provisions regarding the burden of proof of the source of work-related accidents, diseases and deaths (Ministry of Labor and Employment, MTE, 2007). This change shifted the focus of Social Security assistance from the individual to the employer. Thus, when a worker suffers a work-related injury or disease, it will be statistically linked to the professional activity, and the Social Security agency will automatically grant the proper work-related accident benefit. Furthermore, the frequency of diseases and accidents resulting from working conditions is associated with a profile of certain professional categories.

The institution of the NTE implies accountability for risk exposure and disease, leaving to the company the burden of proof that these events are not work-related. This legal device also specifies the worker's job security for 12 months after his rehabilitation. In addition, companies must contribute to the Accident Insurance Factor (FAP), on a risk scale of one to three, according to disease and accident occurrence rates, imposing on them the need to change their risk prevention strategies. According to the risk, companies make differentiated Workplace Accident Insurance contributions.

The establishment of the NTE requires the inclusion of information about the International Classification of Diseases (ICD) and the National Classification of Economic Activities (CNAE) of the Ministry of Labor. Thus, even if the company does not give formal notice of an accident, the information is automatically cross-referenced when the employee seeks medical or social security assistance.

Originating from Brazil's Ministry of Health, the proposal of the Worker Health Surveillance (VST) program is based on the model of epidemiological study of injuries at the various levels of the relationship between work and health, adding to the universe of evaluation and analysis the ability for immediate intervention regarding the determining factors of damage to health. The Normative Instruction of Occupational Health Surveillance in the SUS was approved by Order no. 3120/GM/MS of July 1, 1998. The Occupational Health Surveillance program is not an unconnected and independent area of public health surveillance as a whole; instead, it is intended to add to the set of actions of public health surveillance strategies of knowledge production and mechanisms of intervention on production processes, bringing the various common objects of sanitary practices closer to those originating from the relationship between work and health (Ministry of Health, MS, 1998).

The VST program promotes a continuous and systematic action to detect, learn, research, and analyze the determining and conditioning factors of health problems related to work processes and environments, from their technological, social, organizational and epidemiological standpoints. Its purpose it to plan, execute and evaluate interventions that eliminate or control their harmful effects on worker health (Ministry of Health, MS, 1998).

The VST program is operationalized by the National Network for Workers' Inclusive Healthcare – RENAST (Ministry of Labor and Employment, MTE, 2005b), whose proposal is comprehensive healthcare for the worker. This network is composed of State and Regional Occupational Health Reference Centers (CEREST), which, in their areas of action, provide technical support, permanent education, coordination of aid projects, and the promotion and surveillance of worker health.

In the international context, studies of the incidence of ergonomic hazards led the European Agency for Safety and Health at Work (EASHW), in partnership with the ILO, to create a Risk Observatory to anticipate new and emerging risks to occupational safety and health. To identify and anticipate tendencies in Occupational Safety and Health (OSH) in Europe, the Observatory allocates resources which enable more timely and effective interventions. These studies are aimed at underpinning legislation promoting the adaptation of work to man (ILO, 2010).

Based on this context, it is clear that not all activities are subject to the same risks, for some groups are subject to higher or specific risks. The OSHA provides, online and upon request, several specific types of publications and materials for the various areas, which can guide and assist in specific procedures for each occupation (OSHA, 2012b).

According to an OSH study developed by the EASHW for new risk prevention strategies from 2007 to 2012, among the factors that offer risks, the main ones are precarious employment contracts, job insecurity, professional instability and stress, and lack of training to perform highly dangerous activities (ILO, 2010).

Many countries have created regional networks of government institutions, researchers and experts to promote and exchange or share their national strategies, best practices, training, and investigations into emerging risks. Other regions, such as Asia, East Africa, Latin America and the Caribbean, work hand in hand with governments, the scientific community, and with professionals on OSH programs (ILO 2010).

Most of the ILO Member States have committed to implement programs to promote decent work in their countries, many of which emphasize the need to improve OSH for all workers, ensuring better health conditions at work to reduce occupational diseases and accidents. This global approach is crucial with respect to new and emerging risks to safety and health at work, in a changing global economy, to ensure sustainable economic and social development (ILO, 2010).

Final Remarks

Ergonomic studies enable one to evaluate the extent to which the working conditions of healthcare workers pose potential occupational risks, and indicate that there are significant differences according to the various categories. Among these categories, nurses display a high prevalence of health problems, demonstrating that the working conditions of nurses are worse than those of other healthcare professionals. Several recommendations by international organizations reveal concerns about illness among healthcare workers, but these concerns have yet to be incorporated in the labor legislation of different countries.

References

Abdalkader RH, Hayajneh FA. Effect of night shift on nurses working in intensive care units at Jordan University Hospital. *European Journal of Scientific Research,* 2008, 23 (1): 70-86.

Almeida RG. A ergonomia sob a ótica anglo-saxônica e a ótica francesa [Ergonomics from the Anglo-Saxon and French perspectives]. *Vértices,* 2011, 13 (1): 115-126.

Associação Brasileira de Ergonomia [Brazilian Ergonomics Association] - ABERGO. O que é ergonomia? [What is ergonomics]? Available at: http://www.abergo.org.br /internas.php?pg= o_que_e_ergonomia [viewed 28 Jun 2012].

Barboza MCN, Milbrath VM, Bielemann VM, Siqueira HCH. Doenças osteomusculares relacionadas ao trabalho (DORT) e sua associação com a enfermagem ocupacional [Work-related musculoskeletal disorders and their association with occupational nursing]. *Revista Gaúcha de Enfermagem,* 2008, 29 (4): 633-638.

Benevides-Pereira AMT, Moreno-Jiménez B. O burnout e o profissional de psicologia. [Burnout and the psychology professional]. *Revista Eletrônica InterAção Psy.,* 2003, 1 (1): 68-75.

Benevides-Pereira AMT. O processo de adoecer pelo trabalho. [The work-related illness process]. In: Benevides-Pereira AMT (org). Burnout: quando o trabalho ameaça o bem-estar do trabalhador. [Burnout: when work threatens the employee's well-being]. São Paulo: Casa do Psicólogo, 2002: 21-92.

Biehl KA. Burnout em psicólogos [Burnout among psychologists]. Faculdade de Psicologia da Pontifícia Universidade Católica. Rio Grande do Sul, 2009. Doctoral thesis.

Boiselle PM, Levine D, Horwich PJ, Barbaras L, Siegal D, Shillue K, Affeln D. Repetitive stress *Journal of the American College of Radiology,* 2008, 5 (8): 919-923.

Bonzini M, Sarquis LMM, Felli VEA, Carugno M, Pesatori AC, Bertù L. Association between psychosocial factors and musculoskeletal disorders across different cultural backgrounds: a comparative study among Brazilian and Italian nursing personnel. *Journal of Occupational and Environmental Medicine,* 2011, 68: A68.

Borin LHV. Avaliação ergonômica e melhoria das condições de trabalho e do processo produtivo de nutrição parenteral total em farmácia de manipulação: um estudo de caso. [Ergonomic evaluation and improvement of work conditions and of the productive process in nutrition: a case study]. Universidade Federal do Rio Grande do Sul. Porto Alegre, 2004. Master's Dissertation.

Bork BE, Cook TM, Rosecrance JC, Engelhardt KA, Thomason MJ, Wauford RK, Worley RK. Work-related musculoskeletal disorders among physical therapists. *Physical Therapy*, 1996, 76 (8): 827-835.

Campo M, Weiser S, Koenig KL, Nordin M. Work-related musculoskeletal disorders in physical therapists: a prospective cohort study with 1-year follow-up. *Physical Therapy*, 2008, 88 (5): 608-619.

Carayon P. Handbook of human factors and ergonomics in health care and patient safety. 2nd ed. CRC Press: Taylors and Francis Group, 2012.

Célia RCRS, Alexandre NMC. Aspectos ergonômicos e sintomas osteomusculares em um setor de transporte de pacientes. [Ergonomic aspects and musculoskeletal symptoms in the transport sector]. *Revista Gaúcha de Enfermagem*, 2004, 25 (1): 33-43.

Chen H, Schernhammer E, Schwarzschild MA, Ascherio A. A prospective study of night shift work, sleep duration, and risk of Parkinson's disease. *American Journal of Epidemiology,* 2006, 163 (8): 726-730.

Cole DC, Hogg-Johnson S, Manno M, Ibrahim S, Wells RP, Ferrier SE. Reducing musculoskeletal burden through ergonomic program implementation in a large newspaper. *International Archives of Occupational and Environmental Health*, 2006, 80 (2): 98-108.

Concha-Barrientos M, Nelson DI, Driscoll T, Steenland NK, Punnett L, Fingerhut MA. Selected occupational risk factors. In: World Health Organization. Comparative Quantification of Health Risks. Global and Regional Burden of Disease Attribution to Selected Major Risk Factors. Geneva, 2005, chapter 21. Available at http://www.who.int/publications/cra/chapters/volume2/1651-1802.pdf [viewed 30 Jun 2012].

Costa FOC, Pietrobon L, Fadel MAV, Regis Filho GI. Doenças de caráter ocupacional em cirurgiões-dentistas: uma revisão de literatura. [Occupational diseases in dental surgeons: a literature review]. In: XXVI Encontro Nacional de Engenharia de Produção ENGEP, 2006, Fortaleza, Brazil.

Covolan MA. Stress ocupacional do psicólogo clínico: seus sintomas, suas fontes e as estratégias utilizadas para controlá-lo. [Occupational stress of clinical psychologists and their symptoms, causes and control strategies]. In Lipp MEN (org). Pesquisas sobre stress no Brasil: saúde, ocupações e grupos de risco. [Research about stress in Brazil: health, jobs and risk groups]. Papirus, 1996: 225-40.

Cromie JE, Robertson VJ, Best MO. Work related musculoskeletal disorder and culture of physical therapy. *Journal of Physio Therapy*, 2000, 82 (5): 459-472.

Dembe AE. The social consequences of occupational injuries and illnesses. *American Journal of Industrial Medicine*, 2001, 40: 403-417.

Directorate for Health and Social Affairs of Norway. A National Clinical Guideline for the Use of Dental Filling Materials: information for dental health care personnel. Norway, 2003. Available at: http://helsedirektoratet.no/publikasjoner/national-clinical-guideline-for-the-use-of-dental-filling-materials [viewed 20 Jun 2012].

Edlich RF, Rhoads SK, Cantrell HS, Azavedo SM, Newkirk AT. Banning mercury amalgam. Food and Drug Administration, Medical Devices Advisory Committee, 2009. Available at: http://www.fda.gov/downloads/AdvisoryCommittees/CommitteesMeetingMaterials/MedicalDevices/MedicalDevicesAdvisoryCommittee/DentalProductsPanel/UCM236379.pdf [viewed 20 Jun 2012].

Eriksen W. The prevalence of musculoskeletal pain in Norwegian nurses aides. *International Archives of Occupational and Environment Health*, 2003, 76 (8): 625-630.

Faber B. Clinical psychologists' perceptions of psychotherapeutic work. *Clinical Psychologist*, 1985, 38 (1): 10-13.

Felli VEA et al. Prevalence of musculoskeletal pain and associated disability in three cohorts of Brazilian workers. In: 22nd International Conference on Epidemiology in Occupational Health. Book of Abstracts, Oxford, 2011.

Food and Drug Administration. Medical advices: products and medical procedures – dental amalgam. Online content. Available at: http://www.fda.gov/MedicalDevices/ProductsandMedicalProcedures/DentalProducts/DentalAmalgam/ucm171094.htm [viewed 30 Jun 2012].

Gaidzinski RR. Dimensionamento do pessoal de enfermagem. [Nursing staff downsizing]. In: Kurcgant P (coord). Administração em enfermagem. [Nursing administration]. EPU, São Paulo, 1991: 91-97.

Garbin AJI, Garbin CAS, Diniz DC. Normas e diretrizes ergonômicas em odontologia: o caminho para a adoção de uma postura de trabalho saudável. [Ergonomic standards and guidelines in dentistry: the path to the adoption of a healthy sitting working posture]. *Revista de Odontologia da Universidade Cidade de São Paulo*, 2009, 21 (2): 155-161.

Gjesdal S, Bratberg E, Mæland JG. Gender differences in disability after sickness absence with musculoskeletal disorders: five-year prospective study of 37,942 women and 26,307 men. *BMC Musculoskeletal Disorders,* 2011, 12: 37.

Gomes AR, Cruz JF. A experiência de estresse e “burnout”em psicólogos portugueses: um estudo sobre as diferenças de gênero. [Burnout and stress experienced by Portuguese psychologists: a study of gender differences]. *Psicologia: teoria, investigação e prática,* 2004, 9: 193-212.

Gonçalves RBM. Tecnologia e organização social das práticas de saúde: características tecnológicas do processo de trabalho na rede estadual de centros de saúde de São Paulo. [Technology and social organization of health practices: technological characteristics of the work process in a state-run network of health centers in São Paulo]. Hucitec/ABRASCO, São Paulo, 1994.

Grigoletto JC, Oliveira AS, Muñoz SIS, Alberguini LBA, Takayanagui AMM. Exposição ocupacional por uso de mercúrio em odontologia: uma revisão bibliográfica. [Occupational risk due to the use of mercury in dentistry: a bibliographic review]. *Ciência e Saúde Coletiva,* 2008, 13 (2):533-542.

Healy KA, Pak RW, Cleary RC, Colon-Herdman A, Bagley DH. Hand problems among endourologists. *Journal of Endourology*, 2011, 25 (12): 1915-1920.

Hemp P. Presenteeism: at work but out of it. *Harvard Business Review*, 2004, 82 (10): 49-58.

Hinze H, Deleon C, Mitchel WC. Dentists at High Risk for Hearing Loss: protection with custom earplugs. *General Dentistry*, 1999, 47 (6): 600-603.

Holder NL, Clark HÁ, DiBlasio JM, Hughes CL, Scherpf JW, Harding L, Shepard KF. Cause, prevalence and response to occupational musculoskeletal injuries reported by physical therapists and physical therapist assistants. *Physical Therapy*, 1999, 79 (7): 642-52.

International Labor Organization (ILO). Riscos emergentes e novas formas de prevenção num mundo de trabalho em mudança. [Emerging risks and new ways of prevention in a world of work changes]. Transl. AP Portugal, Abril, 2010.

Karwowsky W. New perspectives on human factors in the design and management of advanced manufacturing systems: a review. *The Japanese Journal of Ergonomics*, 1991, 27 (6), 301-312.

Karino M. As causas de mortes dos enfermeiros: uma revisão sistemática. [Causes of death among nurses]. Escola de Enfermagem da Universidade de São Paulo. São Paulo, 2012. Doctoral Thesis.

Keller SD. Quantifying social consequences of occupational injuries and illnesses: state of the art and research agenda. *American Journal of Industrial Medicine,* 2001, 40: 438-451.

Kessler RC, Ames M, Hymel PA, Loepkke R, McKenas DK, Richling D, Stang PE, Üstün TB. Using the WHO Health and Work Performance Questionnaire (HPQ) to evaluate the indirect workplace costs of illness. *Journal of Occupational and Environmental Medicine*, 2004, 46 (suppl.6): S23-S37.

Labarthe N. Pereira MEC. Biossegurança na experimentação e na clínica veterinária pequenos animais [Biosafety in experimentation and in small animal veterinary clinics]. *Ciência Veterinária nos Trópicos*, 2008, 11 (suppl 1): p.153-157.

Larese F, Fiorito A. Musculoskeletal disorders in hospital nurses: a comparison of two hospitals. *Ergonomics,* 2007, 37 (7):1205-1211.

Leite HR, Nunes APN, Couto JMM, Bueno DAA, Alcântara M. Estudo dos sintomas osteomusculares em fisioterapeutas: estudo de caso. [Studies of musculoskeletal symptoms in physiotherapists: a case study]. EFDEPORTES Revista Digital, 2010, 15 (146). Available at: http://www.efdeportes.com/efd146/sintomas-osteomusculares-em-fisioterapeutas.htm [viewed 30 Jun 2012].

Leite PC, Silva A, Merighi MAB. A mulher trabalhadora de enfermagem e os distúrbios osteomusculares relacionados ao trabalho. [Female nurses and the musculoskeletal disorders related to their work]. *Revista da Escola de Enfermagem da USP*, 2007; 41 (2): 287-291.

Lloyd C, Waghnorn G, McHugh C. Musculoskeletal disorders and co-morbid depression: implications for practice. *Australian Occupational Therapy Journal,* 2008, 55: 23-29.

Loiola RA ; Felli VEA. Musculoskeletal disorders among nursing workers and the quality of life. *Ergonomia: An International Journal of Ergonomics and Human Factors,* 2007, 29: 251-256.

Lopes AC, Melo ADP, Santos CC. Estudo dos limiares de audibilidade nas altas frequências em trabalhadores da área odontológica [A study of the high-frequency hearing thresholds of dental professionals]. *International Archives of Otorhinolaryngology,* 2012, 16 (2): 226-231.

Lucas M, Day L, Shirangi A, Fritschi L. Significant injuries *Occupational Medicine (London),* 2009, 59 (5): 327-333.

Magnago TSBS, Lisboa MTL, Griep RH. Estresse, aspectos psicossociais do trabalho e distúrbios musculoesqueléticos em trabalhadores de enfermagem. [Stress, psychosocial aspects of work and musculoskeletal disorders in nursing workers]. *Revista de Enfermagem da UERJ,* 2009; 17 (1): 118-123.

Magnavita N. Perceived work-related strain, anxiety *Giornale Italaliano di Medicina del Lavoro ed Ergonomia*, 2009, 31 (1Suppl A): A24-29.

Mastrangelo G, Perticaroli S, Camipo G, Priolo G, Leva A, de Merich D, Marangi G, Fedeli U, Fadda E, Scoizzato L, Marchiori L. Working and health *Medicina del Lavoro,* 2008, 99 (Suppl 1): 9-30.

McMahon M, Stiller K, Trott P. The prevalence of thumb problems in Australian physiotherapists is high: an observational study. *Australian Journal of Physiotherapy*, 2006, 52: 287-292.

Mierzejewski M, Kumar S. Prevalence of low back pain among physical therapists in Edmonton, Canada. *Disability Rehabilitation*, 1997, 19 (8):309-317.

Mininel VA. Adaptação transcultural do Work Disability Diagnosis Interview (WoDDI) para o contexto brasileiro. [Cross cultural adaptation of the Work Disability Diagnosis Interview (WoDDI) to a Brazilian context]. Escola de Enfermagem da Universidade de São Paulo. São Paulo, 2010. Thesis (PhD).

MTE - Ministério do Trabalho e do Emprego. Lei 8.213 de 24 Julho de 1991. Dispõe sobre os Planos de Benefícios da Previdência Social e dá outras providências. [Ministry of Social Security. Law 8.213, 24 Jul 1991. Rules on social security benefits and other issues]. Federal Official Gazette of Brazil, Brasília, 25 Sep 1991. Sections V and VI, Brazil.

MS - Ministério da Saúde. Portaria n. 3.120 de 1 de julho de 1998. Aprova a Instrução Normativa de Vigilância em Saúde do Trabalhador no SUS. [Ministry of Health. Ordinance 3.120, 01 Jul 1998. Approval of the Regulatory Standard for worker health in the National Health Service –SUS]. Federal Official Gazette of Brazil, Brasília, 7 Jul 1998: section 1: 36, Brazil.

MTE - Ministério do Trabalho e Emprego. Fundacentro: Pontos de Verificação Ergonômica – Soluções Práticas e de Fácil aplicação para melhorar a segurança, a saúde e as condições de trabalho. [Ministry of Labor and Employment. Fundacentro: Ergonomic checkpoints – practical and easy-to-implement solutions for improving safety, health and working conditions]. Prepared by ILO (International Labor Office) in collaboration with the International Ergonomics Association. São Paulo, 2001a. Available at: <http//www.fundacentro.gov.br/domínios/CTN> [viewed 24 Jun 2012], Brazil.

MTE - Ministério do Trabalho e Emprego. Portaria GM n. 3.214, de 08 de junho de 1978. Dispõe sobre as Normas Regulamentadoras em saúde, segurança e medicina ocupacional. [Ministry of Labor and Employment. Ordinance 3.214, 8 Jun 1978: Regulatory Standards for occupational health, safety and medicine]. NR 15 – Insalubrious activities and operations. In: Handbook of Legislation. Atlas of Occupational Safety and Medicine, 49th ed., São Paulo: Atlas, 2001b, Brasil.

MTE - Ministério do Trabalho e Emprego. Manual de aplicação da Norma Regulamentadora n 17 [Ministry of Labor and Employment. Handbook of Regulatory Standard 17]. 2nd ed. MTE, 2002, Brazil.

MTE- Ministério do Trabalho e do Emprego. Norma Regulamentadora NR32. Dispõe sobre a segurança e saúde no trabalho em estabelecimentos de assistência à saúde. [Ministry of Labor and Employment. Regulatory Standard 32 (NR 32). Rules on workplace safety and health in healthcare establishments]. Brasília, 2005a. Available at: http://www.mte.gov.br/legislacao/normas_regulamentadoras/nr_32.pdf [viewed 2 Jul 2012], Brazil.

MS - Ministério da Saúde. Portaria n° 2437 de 07 de dezembro de 2005b. Dispõe sobre a ampliação e o fortalecimento da Rede Nacional de Atenção Integral à Saúde do Trabalhador – RENAST no Sistema Único de Saúde – SUS e dá outras providências. [Ministry of Health. Ordinance 2.437, 07 Dec 2005b. Rules on the extension and strengthening of the National Network for Workers' Inclusive Healthcare – RENAST in Brazil's National Health Service (SUS) and other issues]. 2005b Available at: http://www.registro.sp.gov.br/cerest/arquivos/portarias/PORTARIA_2437.pdf [viewed 5 Jul 2012], Brazil.

MTE - Ministério do Trabalho e do Emprego [Ministry of Social Security]. Decreto-Lei n.6042, de 12 de fevereiro de 2007. Altera o Regulamento da Previdência Social, aprovado pelo Decreto nº 3.048, de 6 de maio de 1999, disciplina a aplicação, acompanhamento e avaliação do Fator Acidentário de Prevenção – FAP e do Nexo Técnico Epidemiológico, e dá outras providências. [Law 6.042, 12 Feb 2007. Changes in the Rules of the Social Security System. Approved by Decree n. 3.048, 6 May 1999, concerning the use, follow-up and evaluation of the Epidemiologic and Technical

Security System Nexus and other issues]. Available at: <http//www.mpas.gov.br> [viewed 24 Jun 2012], Brazil.

MTE - Ministério do Trabalho e Emprego. Lei 6.514, de 22 de dezembro de 1977. Altera o Capítulo V do Título II da Consolidação das Leis do Trabalho, relativo à Segurança e Medicina do Trabalho. Manuais de Legislação Atlas: Segurança e Medicina do Trabalho. [Law 6.514, 22 Dec 1977. Changes Chapter V, item II of the Consolidation of Labor Laws concerning Occupational Medicine and Safety. Handbook of Legislation]. Atlas, Occupational Medicine and Safety. São Paulo, 2010, Brazil.

Montmollin MA. Ergonomy. Lisboa. Instituto Piaget, 1990.

Murofuse NT, Marziale MHP. Doenças do sistema osteomuscular em trabalhadores de enfermagem. [Diseases of the musculoskeletal system among nurses]. *Revista Latino Americana de Enfermagem*, 2005, 13 (3): 364-373.

Nelson DI, Concha-Barrientos M, Driscoll T, Steenland K, Fingerhut M, Punnett L. The global burden of selected occupational disease and injury risks: methodology and summary. *American Journal of Industrial Medicine,* 2005, 48 (6): 400-418.

Nicot AM. Women more at risk for musculoskeletal disorders. ANACT. France, 2008.

Nita ME, Secoli SR, Nobre MRC, Ono-Nita SK, Campino AC, Santi FM. Avaliação de tecnologias em saúde: evidência clínica, análise econômica e análise de decisão. [Evaluation of technologies in health: clinical evidence, economic analysis and decision analysis]. Artmed, Porto Alegre, 2010.

Nogueira SA, Bastos LF, Costa ICC. Riscos ocupacionais em odontologia: revisão de literatura. [Occupational risks in dentistry: a literature review]. *UNOPAR Científica - Ciências Biológicas e da Saúde,* 2010, 12 (3):11-20.

Nordin NAM, Leonard JH, Thye NC. Work-related injuries among physiotherapists in public hospitals – a Southeast Asia Picture. *Clinics,* 2011, 66 (3):373-378.

Nyland LJ, Grimmer KA. Is undergraduate physiotherapy study a risk factor for low back pain (LBP)? *BMC Musculoskeletal Disorder*, 2003, 4: (22). Available at: http://www.ncbi.nlm.nih.gov/pmc/articles/PMC270026/pdf/1471-2474-4-22.pdf [viewed 30 Jun 2012].

Nyssen AS, Côte V. Motivational mechanisms at the origin of control task violations: An analytical case study *Ergonomics,* 2010, 53 (9): 1076-1084.

Obembe AO, Onigbinde AT, Johnson OE, Emechete AAI, Oyinlola MJ. Occupational injuries among physical therapists in South-West Nigeria. *Nigerian Journal of Medical Rehabilitation*, 2008, 13 (1-2): 25-30.

Occupational Safety and Health Administration (OSHAa). Health Care: Risks for health care workers. Online content. Available at: http://osha.europa.eu/en/faq/what-are-the-main-occupational-health-and-safety 30 Jun 2012].

Occupational Safety and Health Administration (OSHAb). US Department of Labor. OSHA Publications and Educational Materials. Online content. Available at: http://www.osha.gov/Publications/OSHA-pubs-listing.pdf [viewed 24 Jun 2012].

Oliver S. Oxford Handbook of musculoskeletal nursing. Oxford University Press, Oxford, 2009.

Pan American Health Organization - PAHO. Estudio comparativo de las condiciones del trabajo y la salut dos trabajadores de la salut em Argentina, Brasil, Costa Rica y Peru. [Comparative study of work conditions and workers health in Argentina, Brazil, Costa Rica and Peru]. Washington, 2012.

Palmer K, Walker-Bone K, Linaker C, Reading I, Kellingray S, Coggon D, Cooper C. The Southampton examination schedule for the diagnosis of musculoskeletal disorders of the upper limb. *Annals of the Rheumatic Diseases,* 2000, 59: 5–11.

Pan A, Schernhammer ES, Sun Q, Hu FB. Rotating Night Shift Work and Risk of Type 2 Diabetes: Two Prospective Cohort Studies in Women. *PLoS Medicine,* 2011, 8 (12): e1001141.

Passier L, McPhail S. Work related musculoskeletal disorders among therapists in physically demanding roles: qualitative analysis of risk factors and strategies for prevention. BMC Musculoskeletal Disorder, 2011, 12 (24). Available at: http://www.ncbi.nlm.nih.gov/pmc/articles/PMC3038991/ [viewed 30 Jun 2012].

Pivetta AD, Jacques MA, Agne JE, Lopes LF. Prevalência de distúrbios osteomusculares relacionados ao trabalho em fisioterapeutas. [Prevalence of work-related musculoskeletal disorders in physiotherapists]. EFDEPORTES Revista Digital, 2005, 10 (80). Available at: http://www.efdeportes.com/efd80/dort.htm [viewed 30 Jun 2012].

Punnett L et al. Estimating the global burden of low back pain attributable to combined occupational exposures. *American Journal of Industrial Medicine,* 2005, 48 (6): 459-569.

Rabin S, Feldman D, Kaplan Z. Stress and intervention strategies in mental health professionals. *British Journal of Medical Psychology,* 1999, 72 (2): 159-169.

Romani JCP. Distúrbios músculo-esqueléticos em fisioterapeutas: incidência, causas e alterações na rotina de trabalho [Musculoskeletal disorders in physiotherapists: incidence, causes and changes in work routine]. Universidade Federal de Santa Catarina. Florianopolis, 2001. Dissertation (Master).

Rossi D, Bravo ES, Takahashi MAC, Maeno M, Cardillo M, Watanabe M. Reabilitação professional pública, um direito do cidadão [Public professional rehabilitation: a citizen's right]. Document for public discussion, 2007. Available at: http: //www.spbancarios.com.br/download/17/proposta_rp_out2007.pdf [viewed 30 Jun 2012].

Roupa Z, Vassilopoulos A, Sotiropoulou P, Makrinika E, Noula M, Faros E, Marvaki C. The problem of lower back pain in nursing staff and its effect on human activity. *Health Science Journal*, 2008, 2 (4).

Salik Y, Özcan A. Work-related musculoskeletal disorders: a survey of physical therapists in Izmir, Turkey. BMC Musculoskeletal Disorders, 2004, 5 (27). Available at: http://www.ncbi.nlm.nih.gov/pmc/articles/PMC516038/ [viewed 30 Jun 2012].

Sanzono CE, Coelho MEC. Estressores e estratégias de coping em uma amostra de psicólogos clímicos. [Stressors and coping strategies in a sample of clinical psychologists]. *Estudos de Psicologia*, 2007, 24 (2): 227-238.

Sharan D, Ajeesh PS. Injury prevention in physiotherapists: a scientific review. *Work,* 2012, 41: 1855-1859.

Shouldice A. Women's heart health at risk with night shift. *The Journal of Queens's University,* 2011, 139 (18).

Smedley J, Inskip H, Trevelyan F, Buckle P, Cooper C, Coggon D. Risk factors for incident neck and shoulder pain in hospital nurses. *Occupational and Environmental Medicine,* 2003, 60: 864-869.

Smedley J, Egger P, Cooper C, Coggon D. Manual handling activities and risk of low back pain in nurses. *Occupational and Environmental Medicine,* 1995, 52: 160-163.

Souza D'Avila L, Fraga Sousa GA, Sampaio RF. Prevalência de trabalho relacionados com os distúrbios osteomusculares entre fisioterapeutas no sistema hospitalar público de Belo

Horizonte. [Prevalence of work-related musculoskeletal disorders among physiotherapists in the public hospital system of Belo Horizonte]. *Revista Brasileira de Fisioterapia*, 2005, 9 (2): 219-225.

Souza HMMR. Ruído, o inimigo invisível: a consideração de um dentista. [Noise, the invisible enemy: a dentist's consideration]. *Revista Brasileira de Odontologia,* 1997, 54 (2):97-101.

Souza JN, Bertolozzi MR. A vulnerabilidade dos trabalhadores de enfermagem frente a tuberculose em um hospital de ensino. [The vulnerability of nursing workers to tuberculosis in a teaching hospital]. *Revista Latino Americana de Enfermagem*, 2007, 5 (2): 259-266.

Surawera IK, Hoe VCW, Kelsall HL, Urquhart DM, Sim MR. Physical and psychosocial factors associated with wrist or hand pain among Australian hospital-based nurses. Injuryprev, on line first, 2011.

Tinubu BMS, Chidozie Mbada E, Oyeyemi A, Fabunmi AA. Work-related musculoskeletal disorders among nurses in Ibadan, South-West Nigeria: a cross-sectional survey. *BMC Musculoskeletal Disorders*, 2010, 11:12.

Trinkoff AM, Lipscomb JA, Geiger-Brown J, Storr CL. Perceived physical demands and reported musculoskeletal problems in registered nurses. *American Journal of Preventive Medicine*, 2003, 24 (3): 270-275.

Umann J, Guido LA, Grazziano ES. Presenteísmo em enfermeiros hospitalares. [Presenteeism among hospital nurses]. *Revista Latino Americana de Enfermagem*, 2012, 20 (1): 159-166.

Varela CDS, Ferreira SL. Perfil das trabalhadoras de enfermagem com diagnóstico de LER/DORT em Salvador-Bahia 1998-2002. [Profile of nurses diagnosed with RSI/WRMSD in Salvador-Bahia in 1998-2002]. *Revista Brasileira de Enfermagem*, 2004, 57 (3): 321-325.

Vidal MC. Introdução à Ergonomia [Introduction to ergonomy]. Rio de Janeiro: Curso de Especialização Superior em Ergonomia, 2011. Available at: http://ebookbrowse.com /introducao-a-ergonomia-vidal-ceserg-pdf-d27748279 [viewed 30 Jun 2012].

Vilar RMS. Produção do conhecimento em ergonomia em enfermagem [Knowledge production in ergonomics and nursing]. Universidade Federal de Santa Catarina. Florianópolis, 2002. Master's Dissertation.

Wahlstedt K, Norbäck D, Wieslander G, Skoglund L, Runeson R. Psychosocial and ergonomic factors, and their relation to musculoskeletal International *Journal of Occupational Safety and Ergonomics,* 2010, 16 (3): 311-321.

World Health Organization – WHO. Global health risks: mortality and burden of disease attributable to selected major risks. Geneva, 2009.

World Health Organization – WHO. Towards a common language for functioning, disability and health ICF. Geneva, 2002. Available at: http://www.fsp.usp.br/~cbcd/ Material/Guia_para_principiantes_CIF_cbcd.pdf [viewed 20 Jun 2012].

Wijnhoven HA, de Vet HC, Picavet HS. Prevalence of musculoskeletal disorders is systematically higher in women than in men. *The Clinical Journal of Pain,* 2006, 22 (8): 717-724.

In: Noise and Ergonomics in the Workplace
Editor: Paulo Henrique Trombetta Zannin

ISBN: 978-1-62417-573-2

Chapter 8

ACOUSTIC EVALUATION OF OPEN PLAN OFFICES: RELATIONSHIP BETWEEN ACOUSTIC PARAMETERS AND USERS' PERCEPTION OF ACOUSTIC QUALITY

Carolina Reich Marcon Passero *
and Paulo Henrique Trombetta Zannin
Laboratory of Environmental and Industrial Acoustics and Acoustic Comfort (LAAICA),
Federal University of Paraná, Brazil

ABSTRACT

Open plan offices are advantageous in that they allow for faster communication among coworkers and are also space savers. However, their disadvantage lies in the fact that they reduce the ability to concentrate and the privacy of the people that work in them. This chapter discusses objective and subjective assessments of the workplace quality of five real open plan offices. The objective evaluations were based on the following measured and simulated acoustic parameters: sound pressure level – STI, reverberation time – RT, speech transmission index – STI, and the rate of spatial decay of sound pressure levels per distance doubling. The subjective evaluations were based on questionnaires answered by the occupants of offices to assess the need for concentration and interaction with colleagues, the most annoying sources of noise, and the requirements of greater privacy and higher speech intelligibility. A comparison of the objective and subjective data revealed a relationship between the responses of the occupants, particularly with regard to the requirements of privacy and speech intelligibility, and the values of the measured and simulated acoustic parameters.

---

* Email: cal.marcon@gmail.com

1. Introduction

Workplaces can be divided into two main environments in terms of acoustic quality: the industrial environment and the business or commercial environment. These environments are divided into various sub-environments according to the activities performed, each with its specific needs. According to Duboc (1998), concern about conditions conducive to the worker's health should start in the design of the workplace. Noise in industrial environments has been widely studied and monitored because they generate very high levels of noise, and workers exposed continuously to such sound pressure levels (SPL) tend to present serious work-related health problems, such as acquired deafness. Although occupational deafness, which is caused mainly by poor quality of the environment and lack of hearing protection for workers, is a problem that deserves attention, the acoustic environment of offices should also be studied (Cordeiro, 1996). Environments with lower noise levels than those found on the factory floor, i.e., the offices of businesses, banks and other commercial enterprises should be treated acoustically, since most intellectual activity is performed in offices. The relative amount of open office space is still increasing and cognitive tasks are performed in these offices. Therefore, efforts should focus increasingly on the acoustic design of open plan offices (Helenius et al., 2007).

2. Open Plan Offices

With regard to types of spatial organization, offices can be divided into two groups: closed or conventional offices, and open plan offices.

According to Durval et al. (2002), open plan offices have been popular among designers since the 1960s and emerged in response to the need for communication and the intense pursuit of productivity. Duffy (1980) stated that the characteristics of offices designed according to this concept are large open plans arranged freely rather than in a geometric or rectilinear manner.

While this form of organizing workspace offers its users certain advantages due to the speed of communication between coworkers, it also leads to loss of comfort, privacy and the ability to concentrate. Therefore, the correct treatment of the acoustic environment is necessary. According to Nijs et al. (2002), the need for acoustic planning of these workspaces has increased, since a large number of environments have been conceived architecturally based on the open layout design.

The absence of high partitions and doors enables viewing the entire office space, information exchanged among its users, easy access to departments, and greater layout flexibility. However, a negative aspect is that it increases the interference of noise between workstations, resulting in poor speech intelligibility and loss of privacy in these workspaces (Nogueira, 2002).

According to Yoon and Loftness (2002), lack of privacy is reportedly the factor of greatest dissatisfaction among occupants of open plan offices. Pejtersen et al. (2006) affirm that the occupants of open plan offices are more acutely aware of thermal discomfort, poor air quality, and noise. In a study conducted by Helenius et al. (2007) in eight offices, with a total of 633 questionnaires filled out, noise and thermal conditions were reportedly the factors of greatest annoyance in the workplace. The acoustic environment of open plan offices has been found to be more unsatisfactory than that of closed offices.

The problems resulting from the removal of partitions and walls reaching the ceiling, which are responsible for much of the acoustic insulation in the workplace, may diminish satisfaction in the work environment. According to Duffy (1980), the main objective of planning the work environment is to increase worker productivity. However, Duboc (1998) emphasizes that noise leads to lack of concentration, lack of privacy, poor communication, irritability and stress, which contribute to diminish productivity.

3. Noise in the Office

Noise is often cited as the most troublesome factor in the internal office environment. In the 1970s and early 1980s, reports of occupants of open plan offices cited lack of privacy and the nuisance caused by noise produced by coworkers (Pejtersen et al., 2006). Even when objective measurements showed that the background noise of conversation in the room was not very high (approximately 50 dB), this noise was a source of distraction for employees (Van der Voordt, 2004b). The sources of noise identified as the greatest distraction were speech, ringing telephones, and the noise caused by people moving around (Helenius et al., 2007; Hongisto et al., 2007). Talking and laughter close to desks were cited as the sources of noise causing the greatest annoyance in open plan offices (Helenius et al., 2007).

Asselineau (2007) considers the signal to noise ratio an important parameter for assessing the acoustic quality of open plan offices. Occupants usually want not only to be protected from the sound of intrusive and unwanted conversations but also to easily understand the speech of members of their team. Bradley (2007) points out that if background noise levels are very low, the signal to noise ratio will be high and speech privacy will be inadequate. On the other hand, very high background noise levels will provide a high level of speech privacy, but the background noise will intrude, causing the occupants to speak more loudly.

With regard to adequate sound pressure levels (SPL), occupational physicians recommend the following levels as a function of the intellectual demands in the workplace: high demand, $LA_{eq} \leq 50$ dB; intermediate demand, $50 < LA_{eq} < 60$ dB; and remaining cases, $50 < LA_{eq} < 70$ (Bobran, 1995).

Table 1. Possibility of spoken communication and use of the telephone at various background noise levels.

Background noise level [dB(A)]	Need for vocal effort and distance	Possibility of communication	Use of the telephone
55	Normal voice at a distance of 3.00m.	Relaxed communication	Satisfactory
65	Normal voice at 0.90m, loud voice at 1.80m, very loud voice at 3.65m.	Continuous communication	Satisfactory
75	Loud voice at 0.60m, shouting at 2.50m	Intermittent communication	Intermediate
85	Very loud voice at 0.30m, shouting at 0.6-0.9m	Minimal communication	Impossible

Adapted from Cavanaugh and Wilkes (1999).

The Brazilian NBR 10152 (ABNT, 2000) and NR 17 (MTE, 1990) standards establish a limit of LA_{eq}=65 dB for offices. However, the NBR 10152 standard establishes that, for purposes of comfort, this level should reach only up to LA_{eq} =45 dB. Table 1 correlates the possibility of face-to-face and telephone conversation with different levels of background noise.

4. Speech Intelligibility in the Office

Hongisto et al. (2007) claim that it is not the speech sound level that determines its distraction effects but its intelligibility. Speech is the greatest source of noise distraction because it occurs unexpectedly, its volume is variable, and it has the greatest possibility of information content of existing sounds in the office. In a study conducted by Kjellberg (1996), the degree of distraction of workers was more related to sound events, the ability to control noise, and the predictability of noise than the sound levels *per se*. Krajcarz (2007) states that for an office to be considered an acceptable environment by its users, they must be sure that they are being understood and that their conversations remain confidential, which are characteristics that tend to be paradoxical requirements.

When taking physical measurements in open plan offices, Tang and Wong (1998) observed that most of the sound energy was concentrated in the frequency range of the human voice. Moreover, the authors observed a high degree of speech intelligibility in these spaces. Egan (1988) states that speech privacy is influenced by three factors: source, medium and receiver.

According to Standard and 1374-02 (ASTM, 2002), the range of privacy between workstations is determined by the influence that one workplace exerts over the other. This degree of interference depends on the amplitude, direction and orientation of the sound source, attenuation of sound over distance and protection by barriers, sound absorption by the environment, and the SPL at the site of the listener. In addition, the layout of furniture and the speaker's and listener's orientation influence privacy in these environments.

Speech privacy and the disturbance caused by coworkers' speech can be described by the speech intelligibility between workstations. Several laboratory and field studies evaluated only neighboring workstations; however, according to Hongisto et al. (2007), this type of evaluation should encompass the entire Office space, since the nuisance caused by noise is not restricted to the closest workstations. Asselineau (2007) stated that speech intelligibility should be good locally to facilitate conversation among members of the same group. However, this intelligibility should deteriorate with increasing distance.

Speech intelligibility can be evaluated objectively through the speech transmission index (STI). The IEC 60268-16 (IEC, 2011) standard defines the STI as the "physical quantity representing the transmission quality of speech with respect to intelligibility."

Kang (2002) stated that an important characteristic of the STI is that the effects of reverberation, background noise and the contribution of the direction of the sound source, which are normally treated individually, are naturally combined in a single index.

Table 2. Subjective correlation of STI values

Subjective scale	Values of STI
Bad	0.00-0.30
Poor	0.30-0.45
Fair	0.45-0.60
Good	0.60-0.75
Excellent	0.75-1.00

Adapted from the IEC 60268- 16 (IEC, 2011).

Table 3. Subjective perceptions and examples of them in offices, correlating STI values between adjacent workstations

Intelligibility / Privacy / Example in offices	STI
Very bad / Confidential / Between two closed offices with strong insulation between them.	0.00-0.05
Bad / Good / Between two closed offices with normal insulation between them.	0.05-0.20
Poor / Fair / Between workstations in a very noisy open plan office.	0.20-0.40
Fair / Weak / Between desks in a well designed open plan office.	0.40-0.60
Good / Very weak / Between desks in a reasonably well designed open plan office.	0.60-0.75
Excellent / None / Between desks in an open plan office with no acoustic design.	0.75-0.99

Adapted from Hongisto (2005).

The IEC 60268-16 standard relates subjective parameters of speech intelligibility to objective values reached through the STI method. This correlation is shown in Table 2.

For the specific case of offices, considering adjacent workstations, Hongisto (2005) presented a table that relates STI values with intelligibility and speech privacy. This table offers examples of STI values in real offices. This table is reproduced below.

5. Acoustic Evaluation of Open Plan Offices

In studies performed in open plan offices, Hongisto et al. (2007), Virjonen et al. (2009) determined the radius of distraction (r_D) by measuring the STI. The radius of distraction is defined by ISO standard 3382-3:2012 as "distance from the speaker where the speech transmission index falls below 0.50."

In a study conducted by Hongisto et al. (2007), two main descriptors were determined based on measured data for the evaluation of offices:

1. Radius of distraction, r_D,
2. Spatial decay rate of speech, DL_2.

According to these authors, in most cases, DL_2 and r_D suffice to describe the acoustic conditions of an open plan office.

According to ISO standard 14257:2001, the rate of spatial decay of sound pressure levels per distance doubling (DL_2) is the slope, in decibels, per doubling of distance, of the curve of the spatial distribution of sound for a given range of distances. ISO standard 3382-3:2012 defines the parameter of spatial decay rate of speech ($D_{2,S}$) from which DL_2 derives, but using the spectrum of normal speech and A-weighting over the whole frequency range.

The DL_2 describes the amount of sound attenuation in the listener's position at workstations. For Asselineau (2007), the higher the DL_2, the better the acoustic conditions because the noise from activities performed there will more attenuated. Hongisto et al. (2007) state that offices can be organized in an order of acoustic quality according to their DL_2 values. In a study by these authors, the highest DL_2 values were obtained in offices with high sound absorption on both the horizontal and vertical planes, as well as in offices with high wall partitions. The lowest DL_2 values were associated with offices with low wall partitions or without partitions, and low sound absorption on the horizontal or vertical plane.

Chigot (2007) reported that the DL_2, which was originally applied to work environments, has proved to be a consistent and realistic descriptor of the acoustic conditions of open plan offices. According to this author, the DL_2 has been included in standards of acoustic quality of buildings.

The French standard NF S31-080, published in 2006, proposes the use of the DL_2 to evaluate large offices (volumes above 300m³). This standard specifies DL_2 values according to the performance requirements of workers: standard performance, DL_2>2dB(A); efficient performance, DL_2>3dB(A); high performance, DL_2>4dB(A) (Chigot, 2007).

Hongisto et al. (2007) observed that high DL_2 values did not ensure high speech privacy, i.e., low STI values. Table 4 lists preliminary recommendations for DL_2 and r_D based on studies conducted by these authors.

Reports by Hongisto el al. (2007) about their studies in offices included data on reverberation time (RT) and ambient noise levels, as well as DL_2 and r_D. According to ISO standard 3382-3:2012, reverberation time used to be regarded as the predominant indicator of acoustic properties, despite evidence that a more complete evaluation requires a survey of various acoustic parameters such as the rate of spatial decay of sound pressure levels, the speech transmission index and background noise levels.

Beranek (2004) describes RT as the time required for the sound in a given space to become totally inaudible after its source has stopped. The ISO standard 3382-2:2008 defines RT more technically as the time required for the SPL to decrease to 60 dB, which is expressed in seconds.

With regard to the ideal RT in open plan offices, Chigot (2007) presents a list of values of reference for RT established in the national legislation of the following countries: Norway, Finland, Sweden and Australia/New Zealand, as indicated in Table 5.

Table 4. Relationship between the acoustic classification of open plan offices and values of DL_2 and r_D

Acoustic classification	DL_2 (dBA)	r_D (m)
Excellent	> 11	< 5
Good	8 to 11	5 to 8
Fair	5 to 8	8 to 11
Poor	< 5	> 11

Adapted from Hongisto et al. (2007).

Table 5. Ideal values of RT in open plan offices

Country	Norway	Finland	Sweden	Australia/New Zealand
Standard or regulation	NS 8175	SFS 5907	SS 025268	AS/NZS 2107
Year of publication	2005	2004	2001	2000
Expression of the acoustic descriptor	Ideal times in octave bands	Maximum mean value of the room in each octave band	Highest value in each octave band	Maximum mean value of the room in each octave band
Broad frequency (Hz)	125 – 2000	250 – 4000	250 – 4000	500 or 1000
Conditions of the room		Furnished	Furnished and unoccupied	
Variations		±0.1 s variation 125 Hz: ±0.3 s	At 125 Hz: +20% permissible	
RT for rooms with more than 300 m^3 (s)	0.60 – 0.80	0.35 – 0.55	0.40	0.40 – 0.60

Adapted from Chigot (2007).

In addition to the above cited values established by national legislation, the German standard VDI 2569:1990 – Schallschutz und Akustische Gestaltung im Büro (Association of German Engineers – Noise control and acoustic treatment in offices) recommends RT≤ 0.5 s as the mean reverberation time at frequencies of 500, 1000 and 2000 Hz (Freymuth et al. 2002).

The French standard NF S31-080, published in 2006, specifies RT values according to the required performance level of workers (Chigot, 2007). This standard specifies RT≤1.2 s for normal performance requirements, RT≤1.0 s for efficient performance, and RT below or equal to 0.8 s for high performance. These values apply if the building has a volume of more than 250 m^3. However, this standard specifies the preferential use of the DL_2, indicating that the reverberation time should be used for the acoustic evaluation of an office only if the distance between the walls is less than 6m.

The above overview clearly indicates the complexity and importance of studying the acoustic conditions in open plan offices. Therefore, the next section describes architectural and acoustic characteristics of real offices and the perception of the occupants of these offices about the acoustic quality in these spaces.

6. A Case Study in Real Open Plan Offices

This section describes five open plan offices with distinct volumes and architectural characteristics. This heterogeneity is useful to evaluate the influence of the constructive characteristics and volume on the acoustic environment of offices.

The subjective data presented here were obtained from questionnaires applied to occupants. The acoustic measurements were used to calibrate the computational model, while the results of SPL presented here correspond to *in situ* measurements. The results of the acoustic parameters STI, r_D, RT and DL_2 presented here were obtained from computer simulations performed with Odeon 9.0 software.

The *in situ* measurements and computational simulations were performed according to the guidelines of national and international standards (ISO 3382-2:2008, IEC 60268:2011, ISO 14257:2001; and NBR 10151:2000). The method used here is described by Zannin et al. (2009) and Passero and Zannin (2012 a, b).

Figure 3. Overall view of offices 1, 2, 3, 4 and 5, from left to right.

6.1. Objective Evaluation

Table 6 presents the constructive and acoustic characteristics of the evaluated offices.

An analysis of Table 6 indicates that the RT measured in the offices was strongly influenced by the volume of the rooms: rooms with larger volumes presented longer RTs. The RT values of offices with similar volumes, i.e., 3 and 5, were more influenced by the sound absorption of the ceiling material than by that of the flooring material. Office 3 had acoustic lining with sound absorption of 0.49 covering only 11% of the length of the ceiling and carpeting with α=0.31 on 86% of the floor.

Office 5 had acoustic lining with α=0.61 on the entire ceiling and floor carpeting with α=0.31 on 11% of the floor. The mean RT of office 5 was practically 50% lower than that of office 3. These findings confirm the importance of ceiling finishing material in reducing the reverberation time in open plan offices.

As for the r_D values, these were found to be strongly influenced by background noise levels. The lowest r_D was found in office 1, where the highest SPL was measured.

An analysis of the values of DL_2 indicates that the highest values that are suitable for the performance of activities in open plan offices were measured in offices 2 and 4.

Table 6. Constructive and acoustic characteristics of the offices

Characteristics	Office 1	Office 2	Office 3	Office 4	Office 5
Area (m²)	613.03	1324.4	902.38	162.00	808.91
Volume (m³)	1716.48	4922.02	2973.54	437.40	2022,28
No. of workstations	147	242	117	62	109
Density (stations/m²)	0.24	0.18	0.13	0.14	0.13
Main ceiling height (m)	2.80	3.60	2.80	2.70	2.50
Ceiling material (1)	0.49	0.02	67% (2): 0.02 22% (2): 0.10 11% (2): 0.49	0.49	0.61
Flooring material (1)	85% (2): 0.18 15% (2): 0.01	0.18	86% (2): 0.31 14% (2): 0.03	0.03	89% (2): 0.03 11% (2): 0.31
Partition height between workstations (m)	-	1.30	1.15 (3)	1.20 (4)	-
Partition material between workstations (1)	-	0.16	0.16	0.07	-
Sound pressure level (dB(A))	64.1	59.2	56.0	55.4	57.7
RT (s) (5)	0.52	1.08	0.88	0.35	0.46
rD (m)	1.00	1.50	2.00	2.50	2.00
DL2,Co (dB(A))	3.01	3.61	3.08	3.62	3.37

(1) Mean α between the frequencies of 125 to 4000 Hz in octave bands. (2) Percentage of total floor or ceiling area covered with material under the description α. (3) Only at a few workstations; other workstations had no partitions. (4) Partial partition, occupying about 1/3 of the dividing line between the workstations. (5) Average value of the octave band frequencies from 500 to 2000 Hz.

6.2. Subjective Evaluation

The subjective responses about the acoustic conditions in the offices were obtained by means of questionnaires. A total of 404 questionnaires were answered, representing an average of 81 questionnaires per office. The questionnaires contained objective questions, each with a scale of five points, where "0" indicated *none* and "4" indicated *significant*.

The purpose of questions 2 and 3 of the questionnaire was to ascertain the need for concentration and for integration of the users with coworkers in the offices in order to perform their tasks. Table 7 lists the mean value and standard deviation obtained for each question. The higher the mean value of the answers to questions 2 and 3 the greater the need for concentration and integration, from the point of view of the respondents.

As can be seen in Table 7, the mean values of the answers to questions 2 and 3 were similar in all the offices. In offices 1, 2 and 3, the mean value of the response to question 2 was slightly higher than that obtained for question 3, while the opposite was found in offices 4 and 5. Based on these findings, it was concluded that the need for concentration of the occupants in offices 1, 2 and 3 is higher than the need for integration with coworkers, and that the need for concentration is higher in office 2, while the need for integration with coworkers is lower in office 3. For the occupants of offices 4 and 5, integration with coworkers was more important than the need for concentration, with office 4 showing less need for concentration and greater need for integration with coworkers.

However, it should be noted that the occupants of all the offices cited the need for concentration and for integration with coworkers, since the mean value of the responses to questions 2 and 3 was higher than that of the question, which was 2.

Questions 4, 8 and 9 relate to the disturbance caused by noise during work activities and the perception of and satisfaction with the acoustic environment of the offices. Question 4 asked whether noise hindered the performance of tasks in the office. Hence, the lower the mean value obtained in response to this question the lower the interference of noise in the work environment. Question 8 refers to the room noise level. This question is not answered on a scale of numbers, but on a subjective scale of *Very Quiet*, *Quiet*, *Pleasant*, *Noisy*, to *Very Noisy*. A number ranging from 0 to 4 was attributed to analyze each subjective parameter, and the result of the most suitable environment was found to be 2, which corresponds to the attribute *Pleasant*. Question 9 has to do with the acoustic environment and satisfaction with it. On a scale of 0 to 4, the highest value corresponds to an individual perfectly satisfied with the acoustic environment of the office. The results for these questions are given in Table 8.

Table 7. Need for concentration and integration with coworkers – mean and standard deviation (STD) values

Question	Office 1		Office 2		Office 3		Office 4		Office 5	
	mean	STD	mean	STD	mean	STD	mean	STD	mean	STD
2- concentration	**3.26**	0.68	**3.44**	0.70	**3.40**	0.70	**3.02**	0.79	**3.12**	0.70
3- integration	**3.13**	0.77	**2.91**	0.80	**2.72**	0.87	**3.32**	0.77	**3.31**	0.74

Table 8. Disturbance caused by noise, perception of noise and satisfaction with the acoustic environment of the offices – mean and standard deviation (STD) values

Question	Office 1		Office 2		Office 3		Office 4		Office 5	
	mean	STD	mean	STD	mean	STD	mean	STD	mean	STD
4- disturbance	**2.57**	1.13	**3.00**	0.98	**3.07**	0.86	**2.61**	1.10	**3.31**	0.74
8- noise	**3.20**	0.62	**2.99**	0.69	**2.88**	0.54	**2.64**	0.53	**2.50**	0.65
9- satisfaction	**1.76**	1.03	**1.44**	0.98	**1.79**	0.79	**2.05**	1.03	**2.29**	0.91

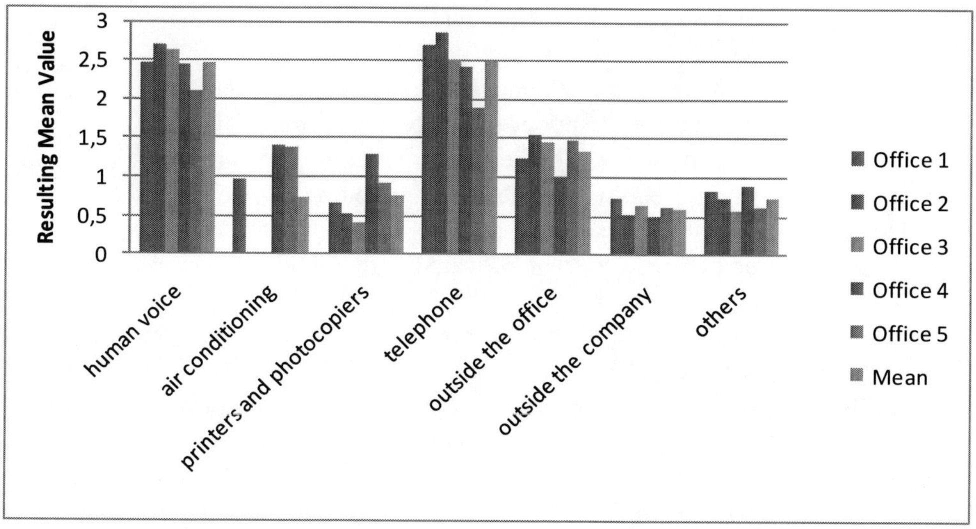

Figure 4. Sources of greatest annoyance in evaluated offices.

An examination of Table 8 indicates that the answers to question 4 reveal that noise is annoying for the occupants of the five evaluated offices. This conclusion is based on the mean values higher than 2. However, also note the high standard deviation of the responses to this question, which showed variations, especially in offices 1 and 4, where the standard deviation was higher than 1. In addition, it can be stated that the occupants of office 5 were the ones most disturbed by noise. The standard deviation was also high for offices 1 and 4, which presented the lowest mean value for this question.

Question 8 elicited a response of close to 3 from the five offices. Thus, it can be stated that the occupants characterized their work environments as noisy, except for office 5, which was characterized as pleasant to noisy (2.50). The noisiest work environment, according to its occupants, was office 1 (3.20). The low standard deviation indicates little variation in the responses.

As for question 9, which referred to the occupants' satisfaction with the acoustic conditions of their work environments, the mean values close to 2 indicated an intermediate degree of satisfaction, i.e., not completely dissatisfied nor totally satisfied. The value of 1.44 for office 2 indicates the least satisfied individuals, while the value of 2.29 for office 5 indicates the most satisfied users among all the work environments analyzed here. The responses obtained from the questionnaires varied considerably, as indicated by the standard deviations of close to 1 calculated for this question.

The purpose of question 5 was to identify the source of greatest disturbance in the five offices. The choice of sources was selected from the literature as normally representing the highest disturbance in open plan offices. The respondents were asked to tick a value from 0 to 4 according to his opinion of the degree of disturbance caused during working hours by each of the six cited noise sources. There was also an option of "others" to verify if the cited sources were suitable to evaluate noise-related disturbance in the work environments, or if there was another specific sound source that would cause greater annoyance than the six listed sources. Figure 4 illustrates the responses to this question.

The mean value of close to 2.5 shown in Figure 4 clearly indicates that the main sources of disturbance in the offices are the human voice and the telephone.

In offices 1 and 2, the main source of annoyance is the sound of ringing telephones, while in offices for 3, 4 and 5 it is the conversations of coworkers (human voices).

The mean value related to the noise caused by air conditioning should not be considered, since offices 2 and 3 were not equipped with air conditioning. However, an analysis of the value obtained for the other offices confirms that the noise produced by air-conditioners is a source of disturbance for the occupants of offices 4 and 5, both of which showed a mean value of 1.4.

Another source of noise that causes some annoyance to users is sounds outside the office, but inside the company, which showed a mean value of 1.3. The respondents considered noises caused by printers and photocopiers and noises outside the company less annoying. Only office 4 reported some annoyance caused by the noise of printers and photocopiers, with a mean value of 1.3.

The value obtained for "others" was lower than 1, indicating that the noise sources listed in this questionnaire encompassed the majority of sources normally found in offices.

Questions 6 and 7 related, respectively, to lack of privacy and to speech intelligibility. Question 6 asked the respondent if he felt annoyed by the fact that his coworkers could listen to his conversations (privacy), while question 7 asked if it was difficult to understand his coworkers when they tried to strike up a conversation with him (intelligibility). The answers to these questions were rated on a scale of 0 to 4. The lower the rate attributed by the respondent the stronger the perception of lack of privacy and speech intelligibility or the lower the demand for them. Table 9 lists the results of the responses to these questions.

An analysis of the values listed in Table 9 clearly indicates the greater demand for privacy than for speech intelligibility in all the offices. The values for privacy were higher than or very close to 2 in all the offices under study.

As for item 7, Lack of intelligibility, most of the offices showed a value close to 1, indicating the perception of good speech intelligibility.

The standard deviations calculated for questions 6 and 7 were quite high, most of them higher than 1. The graphs in Figures 5 and 6 illustrate the respondents' perceptions.

According to the perceptions of most of the respondents in most of the offices, the degree of lack of privacy was between 2 and 3, indicating a greater need for speech privacy in these workplaces. This degree varied from 1 to 2 only in office 4, indicating a fairly high demand for speech privacy.

Table 9. Data on lack of privacy and of speech intelligibility- – mean and standard deviation (STD) values

Question	Office 1		Office 2		Office 3		Office 4		Office 5	
	mean	STD	mean	STD	mean	STD	mean	STD	mean	STD
6- lack of privacy	**1.97**	1.18	**2.16**	1.34	**2.43**	1.20	**2.02**	1.09	**1.93**	1.19
7-lack of intelligibility	**1.67**	1.10	**1.27**	1.06	**1.37**	1.04	**1.14**	0.95	**1.13**	0.86

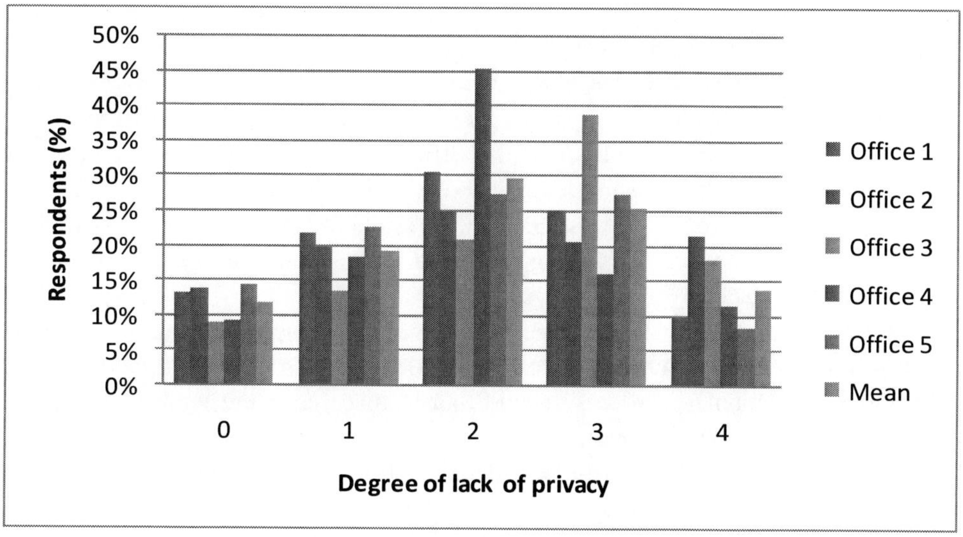

Figure 5. Degree of lack of privacy.

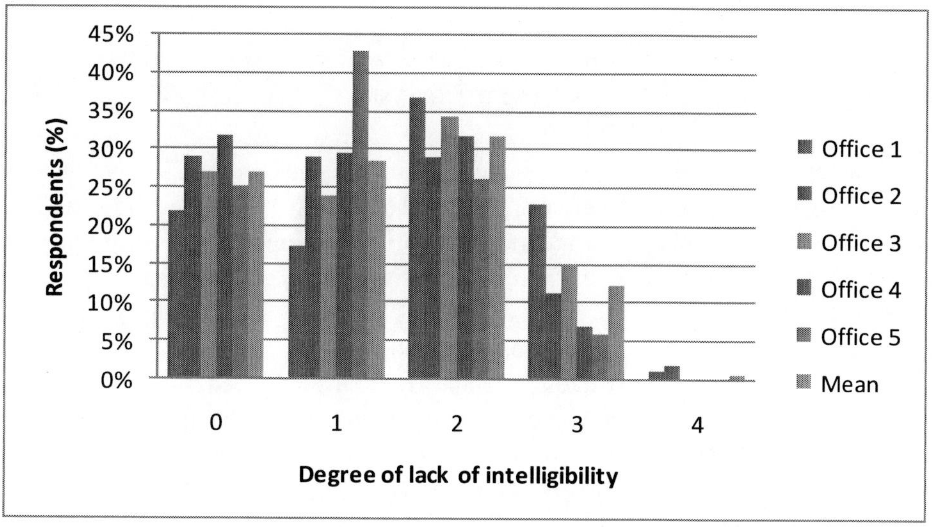

Figure 6. Degree of lack of intelligibility.

Regarding the degree of lack of intelligibility (Figure 6), only 13% rated it at 3 and 4, while 55% assigned a value of 0 or 1 to this parameter and 32% assigned a value 2, which was quantified as the mean value on the scale of 0 to 4. In office 1, a higher percentage of respondents than in the other offices assigned a score of 3 for the degree of lack of intelligibility, indicating the demand for intelligibility in that office.

The responses obtained from the questionnaires thus led to the following conclusions:

1. The users of all the offices stated a strong need for concentration and integration with coworkers to perform their work activities.
2. In offices 1, 2 and 3, the need for concentration was considered more important than the need for integration with coworkers, while the opposite was found in offices 4 and 5.
3. The respondents stated that they were disturbed by noise in performing their activities, and the occupants of office 5 were the most annoyed by noise.
4. Most of the respondents classified the offices as noisy, and only the occupants of office 5 rated it from pleasant to noisy.
5. The occupants were fairly satisfied with the ambient sound level of the offices, with the occupants of office 5 claiming the greatest satisfaction and those of office 2 the least satisfaction.
6. The sources of greatest disturbance were the human voice and ringing telephones. In offices 3, 4 and 5 the human voice was considered the greatest source of annoyance, while the occupants of offices 1 and 2 stated that they were more disturbed by the noise coming from telephones.
7. All the offices showed a higher demand for speech privacy than for speech intelligibility.
8. Only the occupants of office 4 found the speech privacy satisfactory, while the users of the other offices reported a high degree of lack of privacy.
9. Only the occupants of office 1 found speech intelligibility poor, while the users of the other offices stated that it was fair or good.

CONCLUSION

This chapter presented results obtained from objective and subjective acoustic evaluations of five real open plan offices. These assessments were based on questionnaires and on acoustic measurements and simulations of various parameters. Based on these results and on the analysis of the literature, several conclusions were drawn about the acoustics of open plan offices.

With regard to the subjective responses obtained from the questionnaires, it was found that in most of the offices, i.e., offices 1, 2 and 3, the users' need for concentration was greater than the need for integration with coworkers. According to Van der Voordt (2004), the open structure of open plan offices favors greater communication and facilitates the sharing of knowledge and skills among workers. In other words, open plan offices are suitable for activities that require a high level of integration among coworkers. This finding indicates that the real demand of users for integration with coworkers should be determined before deciding whether to design the workspace as either closed or open plan offices.

Most of the respondents classified the offices as noisy. Office 1 was classified as the noisiest workplace, and this classification was confirmed by the objective analysis, which revealed a sound pressure level close to 65 dB(A). According to Bradley (2007) and the Brazilian standard NBR 10152 (ABNT, 2000), the ideal SPL for an office with the characteristics of the workspaces analyzed here is 45 dB(A). However, this level of background noise was not found in any of the five offices, and the lowest SPL, 55.4 dB(A), was found in office 4. According to studies conducted by Evans and Johnson (2000), the noise levels in open plan offices can be highly stressful and demotivating, and according to Hedge (1982), may lead to a high degree of distraction and low level of speech privacy.

The subjective analysis indicated that there was a higher demand for speech privacy than for speech intelligibility in all the offices. According to Yoon and Loftness (2002), lack of privacy is considered the factor of greatest dissatisfaction among the occupants of open plan offices.

An examination of the responses to the questionnaires revealed that only the occupants of office 4 classified speech privacy as fair. The computer simulations of this office indicated that it had the highest r_D among the five evaluated offices, due to the lowest measured SPL. Moreover, simulations of this office showed the highest DL_2 and the lowest RT, indicating high noise reduction over distance, and high noise absorption, and therefore greater speech privacy despite the high r_D.

The respondents of office 1 classified it as offering low speech intelligibility and indicated their need for higher intelligibility. The computer simulations of office 1 showed a low r_D, which in fact was the lowest r_D among all the offices under study, indicating poor speech intelligibility in this workspace. These results clearly illustrate the relationship between the subjective response of demand for speech privacy and intelligibility and the objective analysis based on measured and simulated acoustic parameters.

The subjective analysis revealed that the sources of greatest annoyance in the five offices were human voices and ringing telephones. Similar results were found in studies conducted by Helenius et al. (2007) and Hongisto et al. (2004, 2007), particularly insofar as they pertain to the high degree of annoyance caused by conversations among coworkers. The physical measurements taken in the five offices demonstrated that most of the sound energy was concentrated in the frequencies of 500 and 1000 Hz, which are sounds that typically characterize human speech (Harris, 1998). The same findings were also reported by Tang and Wong (1998).

ACKNOWLEDGMENTS

The authors gratefully acknowledge the German Government, through the German Academic Exchange Service – DAAD (Deutscher Akademischer Austauschdienst) and the Brazilian Government, through the National Council for Scientific and Technological Development – CNPq, for their financial support, which enabled the purchase of the sound level meters and software used in this study.

References

Asselineau M (2007). Integration of Furnishing in Open Plan Office Design: Case Studies. In: International Congress on Acoustics, 19. Madrid. Electronic Proceedings... United Kingdom: Institute of Acoustics, 2007. Available at http://www.sea-acustica.es/WEB_ICA_07/fchrs/papers/rba-10-010.pdf.

Associação Brasileira de Normas Técnicas [*Brazilian Association of Technical Standards*] (ABNT, 2000). NBR 10151: avaliação do ruído em áreas habitadas, visando o conforto da comunidade: procedimento. [*Noise assessment in populated areas to ensure the comfort of the community: Procedure*] Rio de Janeiro, RJ, Brazil (in Portuguese).

Associação Brasileira de Normas Técnicas (ABNT, 2000). NBR 10152: Níveis de ruído para conforto acústico: procedimento. [*Noise levels for acoustic comfort: Procedure*] Rio de Janeiro, RJ, Brazil (in Portuguese).

Beranek L (2004). *Concert Halls and Opera Houses: Music, Acoustics and Architecture.* 2th ed. New York: Springer.

Bobran HW (1995). *Handbuch der Bauphysik.* Stuttgart: völlig neubearbeitete Auflage, 1995.

Bradley JS (2007). Designing and Assessing Speech Privacy in Open-plan Offices. In: International Congress on Acoustics, 19. Madrid. Electronic Proceedings... United Kindom: Institute of Acoustics, 2007. Available at http://www.sea-acustica.es/WEB_ICA_07/fchrs/papers/rba-10-007.pdf.

Cavanaugh W J, Wilkes JA (1999). Architectural Acoustics: Principle and Practice. United States of America: John Wiley and Sons.

Chigot P (2007). Alternative Room Acoustic Descriptors for Open Offices - Progress in Standardization. In: International Congress on Acoustics, 19. Madrid. Proceedings... United Kingdom: Institute of Acoustics, 2007. PACS 43.55.

Cordeiro CVC (1996). Qualidade Acústica em Escritórios Panorâmicos: A Utilização de Sistemas Eletrônicos de Mascaramento. [*Acoustic Quality in Open Plan Offices: The Use of Electronic Sound Masking Systems*] 149 f. Master's Dissertation (Architectural Science) - Federal University of Rio de Janeiro, RJ, Brazil (in Portuguese).

Duboc M (1998). Propagação do Som em Ambiente Panorâmico Caracterizando os Aspectos da Absorção [*Sound Propagation in an Open Plan Environment, Characterizing the Aspects of Absorption*]. 124 f. Master's Dissertation (Mechanical Engineering)- Federal University of Rio de Janeiro, Rio de Janeiro, RJ, Brazil (in Portuguese).

Duffy F (1980). *Oficinas.* Madrid: H. Blume Ediciones.

Durval CL, Charles KE, Veitch JA (2002). *Open-Plan Office Density and Environmental satisfaction.* National Research Council Canada, Canada, IRC Research Report RR-150. Available in: http://irc.nrc-cnrc.gc.ca/pubs/rr/rr150/rr150.pdf..

Egan M D (1988). Architectural Acoustics. New York: McGraw Hill.

Evans GW, Johnson D (2000). Stress and open-office noise. *Journal of Applied Psychology*, Washington, 85 779-783.

Freymuth H.et al. (2002) *Lehrbuch der Bauphysik.* 5. ed Stuttgart: B. G. Teubner.

Harris CM (1998). *Handbook of acoustical measurements and noise control.* USA: Acoustical Society of America.

Hedge A (1982). The open-plan office, a systematic investigation of employees' reactions to their work environment. *Environment and Behavior,* Thousand Oaks, CA, USA, 14 519-542.

Helenius R et al (2007). Acoustic Environment in Finnish Offices: The Summary of Questionnaire Studies. In: International Congress on Acoustics, 19. Madrid. Proceedings... United Kindom: Institute of Acoustics CD-ROM.

Hongisto V, Keranen J, Larm P (2004). Simple model for the acoustical design of open-plan offices. *Acta Acustica,* Stuttgart, 90 481- 495.

Hongisto V, Virjonen P, Keränen J (2007). Determination of Acoustic Conditions in Open Offices and Suggestions for Acoustic Classification. In: International Congress on Acoustics, 19. Madrid. Proceedings...*United Kingdom: Institute of Acoustics,* 2007. CD-ROM.

Hosgito V (2005). A model predicting the effect of speech of varying intelligibility on work performance. *Indoor Air,* USA, 15 (6) 458-468.

International Electrotechnical Commission (2011). IEC 60268- 16: Sound system equipment-Part 16: Objective rating of speech intelligibility by speech transmission index. Switzerland.

International Organization for Standardization (2001). ISO 14257: Acoustics- Measurement and parametric description of spatial sound distribution curves in workrooms for evaluation of their acoustical performance. Switzerland.

International Organization for Standardization (2012). ISO 3382-3: Acoustics — Measurement of room acoustic parameters - Part 3: Open plan offices Switzerland.

International Organization for Standardization (2008). ISO 3382-2: Acoustics-Measurement of room acoustic parameters - Part 2: Reverberation time in ordinary rooms. Switzerland.

International Standards Worldwide (ASTM 2002). E 1374-02: Open Office Acoustics and Applicable ASTM Standards. USA.

Jeon, JY, Lee SC, Vorländer M (2004). Development of scattering for concert halls. *Applied Acoustics,* United Kingdom, 65 341-355.

Kang J (2002). Numerical modeling of speech intelligibility in dining spaces. Applied Acoustics, United Kingdom, 63 1315- 1333.

Kjellberg A et al (1996). The Effects of Nonphysical Noise Characteristics, Ongoing Task and Noise Sensitivity on Annoyance and Distraction due to Noise at Work. *Journal of Environmental Psychology*, USA, 16 123-136.

Krajcarz F (2007). Open Office Acoustic Design: Multi-dimensional Case Studies. In: *International Congress on Acoustics*, 19. Madrid. Proceedings...United Kingdom: Institute of Acoustics, 2007. CD-ROM.

Ministério do Trabalho e Emprego (MTE). Portaria MTPS n.º 3.751, de 23 de novembro de 1990. Norma Regulamentadora: NR-17: Ergonomia. [*Ministry of Labor and Employment (MTE). MTPS Ordinance No. 3751 of November 23, 1990. Regulatory Standard NR-17: Ergonomics*] Brasília, DF, Brazil.

Nijs L et al (2002). *Absorbing Surfaces in Ray-tracing Programs for Coupled Spaces. Applied* Acoustics, United Kingdom, 63 611–626.

Nogueira FF (2002) Análise Paramétrica do Campo Acústico de Escritórios Panorâmicos [*Parametric Analysis of the Acoustic Field of Open Plan Offices*]. 142 f. Master's Dissertation (Civil Engineering)- Federal University of Santa Catarina, Florianópolis, SC, Brazil (in Portuguese).

Passero CRM, Zannin PHT (2012a). Acoustic Evaluation and Adjustment of an Open-Plan Office through Architectural Design and Noise Control. *Applied Ergonomics*, 43 1066-1071.

Passero CRM, Zannin PHT (2012b). Study of the acoustic suitability of an open plan office based on STI and DL2 simulations. *Archives of Acoustics,* 37 237-243.

Pejtersen, J. et al (2006). Indoor climate, psychosocial work environment and symptoms in open-plan offices. *Indoor Air,* USA, 16 (5) 392- 401.

Tang SK, Wong CT (1998). Performance of noise Indices in Office Environment Dominated by Noise from Human Speech. *Applied Acoustics,* United Kingdom, 55 (4) 293-305.

Van der Voordt TJM (2004). Productivity and employee satisfaction in flexible workplaces. *Journal of Corporate Real Estate*, United Kingdom, 6 (2) 133- 148.

Yoon WC, Loftness V (2002). Comparison of two speech privacy measurements, articulation index (AI) and speech privacy noise isolation class (NIC), in open workplaces. *Journal of the Acoustical Society of America, Melville,* NY.111 2441, 2002.

Virjonen P, Keränen J, Hongisto V, (2009). Determination of acoustical conditions in open-plan offices: proposal for new measurement method and target values, *Acta Acustica united with Acustica*, 95 279–290.

Zannin PHT, Passero CRM, Sant'Ana DQ (2009). Acoustic Design of Enclosed Spaces. [in] *Ergonomics: Design, Integration and Implementation.* Brinkerhoff B. [Ed]. New York: Nova Science Publishers.

In: Noise and Ergonomics in the Workplace
Editor: Paulo Henrique Trombetta Zannin
ISBN: 978-1-62417-573-2

Chapter 9

NOISE IN FITNESS GYMS

Karyn Lia Hamad Anjelo*
and Paulo Henrique Trombetta Zannin
Laboratory of Environmental and Industrial Acoustics and
Acoustic Comfort, Federal University of Paraná, Brazil

ABSTRACT

This chapter addresses the problem of noise in fitness gyms. Sound pressure levels were measured and computer simulations were performed to determine the reverberation time, RT, and speech transmission index, STI, in order to characterize the acoustic characteristics of these work and leisure environments. The effects of noise on physical education instructors and clients were identified by means of a questionnaire. The measurements indicated that noise levels reached maximum values, Lmax, of about 108.8 dB(A), which are levels normally found in industrial environments. In view of this finding, the environment of fitness gyms can be described as acoustically unhealthy. An analysis of the responses to the questionnaire revealed that 25% of physical education instructors feel vocal fatigue during gym classes and that this percentage increases to 50% after the classes.

1. INTRODUCTION

Scientific development and technological advances, along with global population growth, contribute markedly to the increase in levels of occupational noise, urban noise (mainly from transport vehicles) and noise produced by a wide range of leisure activities of modern society (fitness gyms, shooting clubs, electronic games, etc.), causing deleterious effects on human health and well-being (Fiorini, 2009).

Noise pollution is present in the daily life of workers, particularly after the Industrial Revolution of the late 18th and early 19th centuries, interfering negatively in the body and causing harmful and often irreversible consequences (Nudelmann et al., 1997).

* E-mail: karyn_adv@yahoo.com.br.

The World Health Organization indicates that, second only to air and water pollution, noise pollution today is the type of pollution that reaches the largest number of people on the planet. Moreover, it is considered a public health problem (WHO, 2003; WHO, 2011). Noise, even at low intensity, affects the human body in various ways, damaging the auditory system, impairing the individual's physical, physiological, mental and social activities, and negatively affecting his performance of work-related activities (Yonezaki and Hidaka, 2005).

Damage caused by noise can be divided into auditory and extra-auditory effects. Auditory changes may be temporary (temporary threshold shift) or permanent, such as acoustic trauma and noise-induced hearing loss (Russo, 1999; Melnick, 1999). A temporary threshold shift is a decrease in the hearing threshold which appears after exposure to intense and continuous noise for a short interval of time, usually accompanied by tinnitus. The hearing threshold is reduced temporarily and returns to normal after the exposure ceases (Russo, 1997; Gonçalves, 2009). Acoustic trauma is referred to as auditory damage due to a single and sudden exposure to very high sound pressure levels caused by impact or impulse noises, which are considered the most harmful to the human ear because they produce irreversible mechanical damage of the cochlea (Russo, 1997; Kós and Kós, 1998; Gonçalves, 2009). NIHL (noise-induced hearing loss) is the result of cumulative noise exposure and is characterized by neurosensory hearing loss, which is usually bilateral and irreversible after it has been established. It occurs primarily at the frequencies of 6 kHz, 4 kHz and 3 kHz, but can extend up to frequencies of 8 kHz, 2 kHz, 1 kHz, 500 Hz and 250 Hz, which aggravate the damage and cause the individual to present intolerance to intense sounds, tinnitus and impaired speech intelligibility (National Committee on Noise and Hearing Conservation, 1994). With regard to organic alterations (extra-auditory effects), the World Health Organization states that noise exposure can cause sleep disorders, cardiovascular diseases and cognitive impairment (WHO, 2011). Other effects are: attention deficit, increased production of thyroid hormones, augmented heart rate, increased production of adrenaline and corticotropin, alterations of the immune system, anxiety, fatigue and diminished work performance (Nudelmann et al., 1997; Penney and Earl, 2004; Kujala et al., 2004; Brattico et al., 2005; Chiovenda et al., 2007).

As for the worker's health, it should be pointed out that work environments offer a variety of environmental and organizational risks that are responsible, along with individual and social aspects, for triggering and increasing the prevalence and incidence of work-related diseases. Note that, among the environmental and occupational health risks, noise is currently considered the most common physical agent in workplaces (Matos, 2008; Gonçalves et al., 2009). According to Costa et al. (2003), high noise levels in Brazil are increasingly related to leisure activities, whether due to excessively loud music or in motor sports and shooting practice. Fitness gyms offer a wide variety of sports activities aimed at improving the quality of life of their patrons. This environment is characterized as occupational for physical education (P.E.) instructors and as a leisure environment for patrons. Notwithstanding the concept of a better quality of life and the pursuit of a healthier life, this environment may also offer health risks to both professionals and users. One of these risks comes from the use of excessively loud music, since, as Maia et al. (2007) point out, although music is pleasurable, it can be harmful to hearing and hence to the quality of life when it is played at high sound pressure levels.

Several studies have demonstrated the use of excessively loud music in fitness gyms, based on measurements of sound pressure levels and the complaints of professionals working in these environments. Investigations have detected sound pressure levels ranging from Leq = 58 to Leq = 105 dB(A), as well as a wide range of effects on health, such as fatigue, malaise, discomfort in response to intense sounds, headache, nervousness, irritability, dizziness, tinnitus, ear fullness, poor concentration, temporary threshold shift and hearing loss. They have also found that personal trainers who have become accustomed to noise are not concerned about the damage that such exposure may cause (Deus and Duarte, 1997; Lacerda, 1999; Pinto and Russo, 2001; Marcon and Zannin, 2004; Palma et al., 2009; Andrade and Russo, 2010). Regarding the use of music in fitness gyms, Zucki and Lacerda (2010) argue that it has become a common practice, since patrons and personal trainers believe it stimulates physical activity, making it more enjoyable and thus enhancing performance. However, many instructors raise the volume of the stereo to very high levels to keep patrons increasingly motivated, thereby impairing their own and their patrons' auditory health (Deus and Duarte, 1997). Several studies conducted in fitness gyms have demonstrated the potential risk of changes in the entire body, especially in the auditory system, due to loud music during physical activity. The findings of these studies indicate that the sound pressure levels in evaluated fitness gyms ranged from 90 to 105 dB(A) in Deus and Duarte's study (1997), from 73.9 to 94.2 dB(A) in Lacerda's study (1999), from 58 to 75.5 dB(A) in Marcon and Zannin's study (2004), and from 74.4 to 101.6 dB(A) in Palma et al.'s study (2009). Complaints such as tinnitus, ear fullness, discomfort in response to intense sounds, headaches, insomnia, intolerance to loud sounds, irritability, nervousness, dizziness, poor concentration, and occasional hearing difficulties were reported by the personal trainers of fitness gyms under study (Deus and Duarte, 1997; Lacerda, 1999; Pinto and Russo, 2001; Andrade et al., 2009; Andrade and Russo, 2010).

Another factor to be considered in this type of environment, and which many researchers believe is an important aspect to analyze, are the acoustic conditions in fitness gyms. The acoustic conditions in these environments include acoustic comfort, which is verified by measuring and/or simulating reverberation times and speech transmission indices, particularly in environments where verbal communication is required (Deus and Duarte, 1997; Palma et al., 2009; Andrade et al., 2009; Zannin et al., 2009; Andrade and Russo, 2010; Packer and Faszer, 2011; Zannin et al, 2011; Zannin et al., 2012).

Given the importance of the theme of leisure activity linked to fitness gyms, the objective of this chapter is to study the acoustic descriptors of these environments – measurement of sound pressure levels, simulations of reverberation time and speech transmission index using three-dimensional models – and to identify the effects of noise on personal trainers and patrons based on their responses to a questionnaire.

2. Measurement and Simulation of Acoustic Descriptors in Fitness Gyms

2.1. Computer Simulation

To evaluate the acoustic quality of fitness gyms, computer simulations were performed to obtain the acoustic descriptors of reverberation time, RT, and speech transmission index, STI, and in situ measurements were taken to determine the equivalent sound pressure levels, Leq.

According to Harris (1994, 1998): "The reverberation time, RT, of a room is the time required for the sound level in the room to decrease 60 dB after the source is stopped." Reverberation time is one of the most important parameters to describe the acoustic quality of rooms – theaters, classrooms, cinemas, concert halls, etc. (Karabiber and Vallet, 2003). Reverberation time depends on: 1) the room's volume; 2) its sound frequency, and 3) its total sound absorption. It is important that the reverberation time be suitable for the intended use of the room. According to the ISO 3382-1 (ISO, 2009) and ISO 3382-2 standards (ISO, 2008), the reverberation time, RT, can be measured by the interrupted noise method and by the integrated impulse response method (Passero and Zannin, 2010). Among the acoustic parameters employed to characterize speech intelligibility, the Speech Transmission Index – STI – is, according to Ronsse (2006): "the most efficient objective metric."

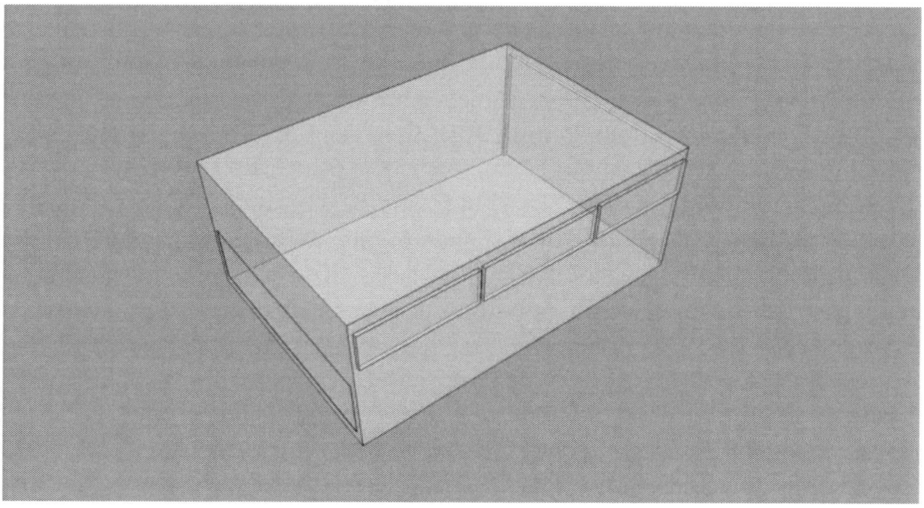

Figure 1. 3D Model of fitness gym 01. Volume = 258.74 m^3.

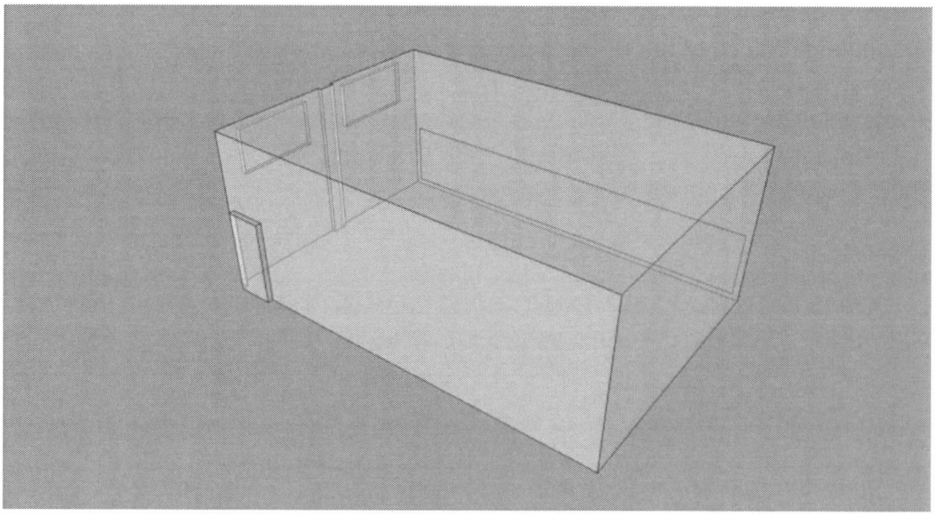

Figure 2. 3D model of fitness gym 02. Volume = 252.84 m^3.

The greater processing capacity of computers, the portability of measuring equipment, and the advent of high speed measurement techniques favor the use of this diagnostic technique. The speech transmission index, STI, is an acoustic descriptor that considers the effects of reverberation, background noise and the contribution of the direction of the source to determine speech intelligibility. These elements, which are usually treated individually, are combined in a single index (Harris, 1998; Kang, 2002). The STI simulations were performed according to the IEC 60268-16 standard (IEC, 2003). To obtain acoustic parameters through simulations required first making a three-dimensional drawing of the room - Figures 1, 2, 3 and 4. Suitable calculation parameters were then inserted (such as the length of the impulse response), the characteristics of the finish surfaces (absorption and scattering coefficients) and the specifications of the sound source and receiver.

For the simulations, finishing materials and their absorption coefficients were attributed to the surfaces of the gym rooms. The calculations were performed using the following ODEON software features: 1) Engineering Method, and 2) Scattering Method. The sound sources used in the simulations were an omnidirectional microphone to calculate the RT and a microphone with directivity similar to the human mouth to calculate the STI. Both sources were simulated at a height of 1.5 m from the floor. The receivers were configured so that each fitness gym had a simple receiver positioned in the middle of the room at a height of 1.2 m from the floor, to obtain the numerical values of the acoustic descriptors RT and STI. A grid was created with receivers placed in a 10x10 cm mesh to ensure the high density of the grid.

Figure 1 shows the three-dimensional model of fitness gym 01, which has the following characteristics: volume = 258.74 m^3, wood paneled ceiling and rubberized flooring. Figure 2 shows the three-dimensional model of fitness gym 02, which has the following characteristics: volume = 252.84 m^3, wood paneled ceiling and ceramic tile flooring. Figure 3 shows the three-dimensional model of fitness gym 03, which has the following characteristics: volume = 169.67 m^3, wood paneled ceiling and wood flooring. Figure 4 shows the three-dimensional model of fitness gym for 04, which has the following characteristics: volume = 393.24 m^3, wood paneled ceiling and wood flooring.

Ananthaganeshan and Gastmeier (2007) recommend a RT of 1.5 to 2 seconds at frequencies of 500 Hz, 1000 Hz and 2000 Hz for gyms. These authors state that a RT within this range makes a compromise between an environment destined for the practice of sports and/or musical performances and speech intelligibility (Ananthaganeshan and Gastmeier, 2007; Packer and Faszer, 2011). Faszer and Pacher (2011) state that according to the Government of Alberta's Design and Construction Standards and Guidelines for School Facilities (2007), the RT in unoccupied gymnasia should not exceed 2 seconds. This value is the arithmetic mean of the RT values obtained at frequencies of 500 Hz, 1000 Hz and 2000 Hz. The IEC 60268-16 standard (IEC, 2003) provides both a quantitative and qualitative assessment of the STI, as indicated in Table 1.

2.2. Measurement of Sound Pressure Levels

The sound pressure levels in the fitness gyms were measured with a class I Brüel and Kjaer 2238 sound level meter.

Measurements were taken of the equivalent continuous sound level Leq and the maximum Lmax and minimum Lmin sound levels.

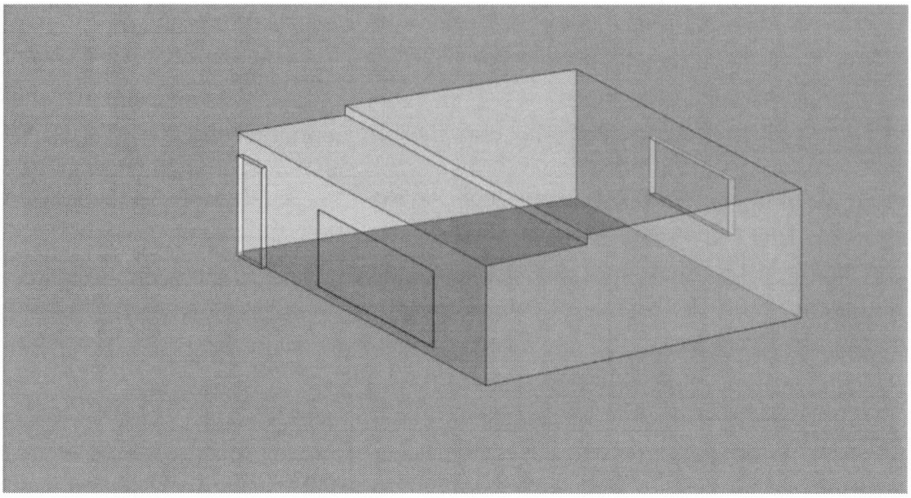

Figure 3. 3D model of fitness gym 03. Volume = 169.67 m^3.

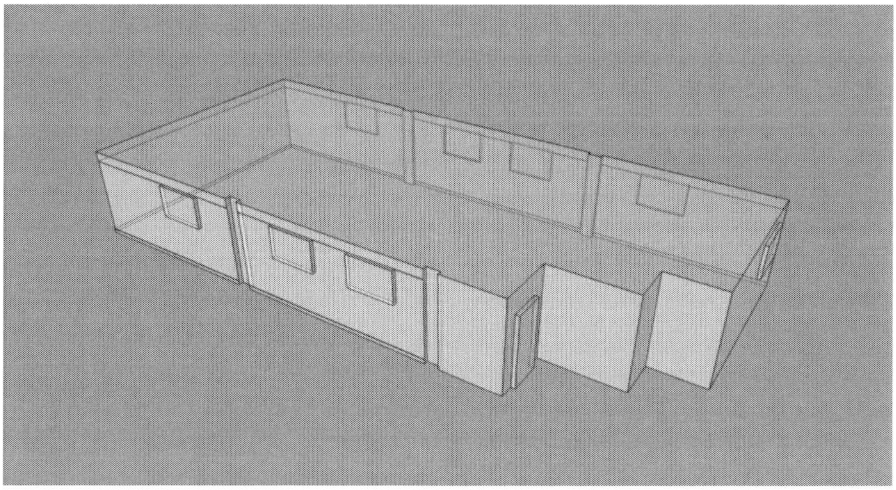

Figure 4. 3D model of fitness gym 04. Volume = 393.24 m^3.

Table 1. Quantitative and qualitative evaluation of intelligibility – IEC 60268-16 (2003)

Quantitative evaluation of intelligibility STI	Qualitative assessment of intelligibility (Subjective impression)
0.75 – 1.00	Excellent
0.60 – 0.75	Good
0.45 – 0.60	Fair
0.30 – 0.45	Poor
0.00 – 0.30	Bad

The measurements were taken for 40 minutes and A-weighted. The sound level meter was placed on a tripod, at a distance of 1.20 m from the walls and at a height of 1.20 m from the floor, as recommended by the Brazilian standard NBR 10151 (2000). The measurements were taken during the following gym workouts: jumping, body balance, localized and aerobic exercise.

2.3. Assessment of Noise Effects

Based on questionnaires as a data collection tool, an investigation was made of the effects of noise on personal trainers and users of fitness gyms. The questionnaires served to garner data on identification, family history, risk factors for hearing loss, workplace environment, leisure habits, and the effects caused by noise. The criterion of inclusion for participation in the sample was to be 18 years or older, while the exclusion criteria were a history of hearing impairment or a family history of hearing deficiency. The sample comprised 4 personal trainers and 21 users. Seven users met the exclusion criteria and were therefore excluded from the original sample.

3. Results and Discussion

The fitness gyms were simulated with ODEON, version 9.2 software, to obtain the acoustic descriptors of RT and the STI index. The average RT was calculated from the simulated data, as the arithmetic mean of the respective reverberation times at frequencies of 500 Hz, 1000 Hz and 2000 Hz (Ananthaganeshan and Gastmeier, 2007). Table 2 describes the RT values and average STI values, as well as the qualitative assessment of STI, as specified in the IEC 60268-16 (2003) standard described in Table 1.Considering the 1.5 to 2 second limit for RT suggested by Ananthaganeshan and Gastmeier (2007) to keep the environment suitable for the practice of physical activity and favorable for music-driven activities, all the fitness gyms evaluated fell within this range except fitness gym 02. The average RT in this fitness gym was 2.6 seconds and the STI indicated Poor speech intelligibility. By computer simulation, the original wood paneled ceiling of fitness gym 02 was replaced with rigid double sided glass wool boards, which have higher sound absorption coefficients.

Table 2. Acoustic descriptors of calculated RT and simulated STI

	Acoustic descriptors for the fitness gyms		Qualitative assessment of intelligibility (see Table 1)
	Calculated RT	Simulated STI	STI
Fitness gym 01	1.8	0.52	Fair
Fitness gym 02	2.6	0.43	Poor
Fitness gym 03	1.6	0.53	Fair
Fitness gym 04	1.9	0.51	Fair

Table 3. Sound absorption coefficients (α) as a function of frequency (Source1: ODEON Software)

Material			Frequencies			
Wood paneling	125 Hz	250 Hz	500 Hz	1 kHz	2 kHz	4 kHz
α[1]	0.08	0.04	0.03	0.03	0.02	0.02

Table 4. Sound absorption coefficients (α) as a function of frequency (Source1: Isover do Brasil)

Material			Frequencies			
Rigid double sided glass wool boards	125 Hz	250 Hz	500 Hz	1 kHz	2 kHz	4 kHz
α[1]	0.20	0.72	0.95	1.10	0.95	0.99

Table 5. Acoustic descriptors of calculated RT and simulated STI for fitness gym 02

	Room with original ceiling		Qualitative assessment of intelligibility (see Table 1)	Room with acoustically treated ceiling		Qualitative assessment of intelligibility (see Table 1)
	RT	STI	STI	RT	STI	STI
Fitness gym 02	2.6	0.43	Poor	1.0	0.63	Good

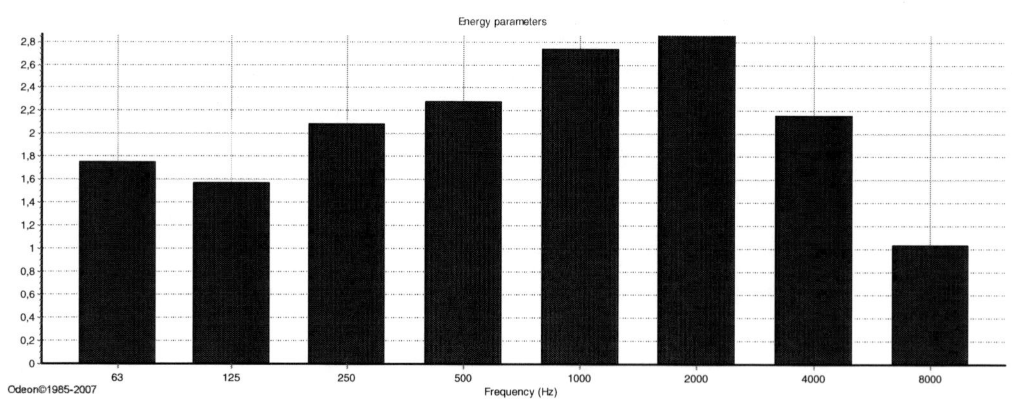

Figure 5. Simulated RT for fitness gym 02 with original ceiling.

Tables 3 and 4 describe the sound absorption coefficients used in the computer simulations for fitness gym 02.

Table 5 shows that the substitution of the ceiling finishing material provided an average RT of 1.0 seconds, while the classification of the STI went from Poor to Good.

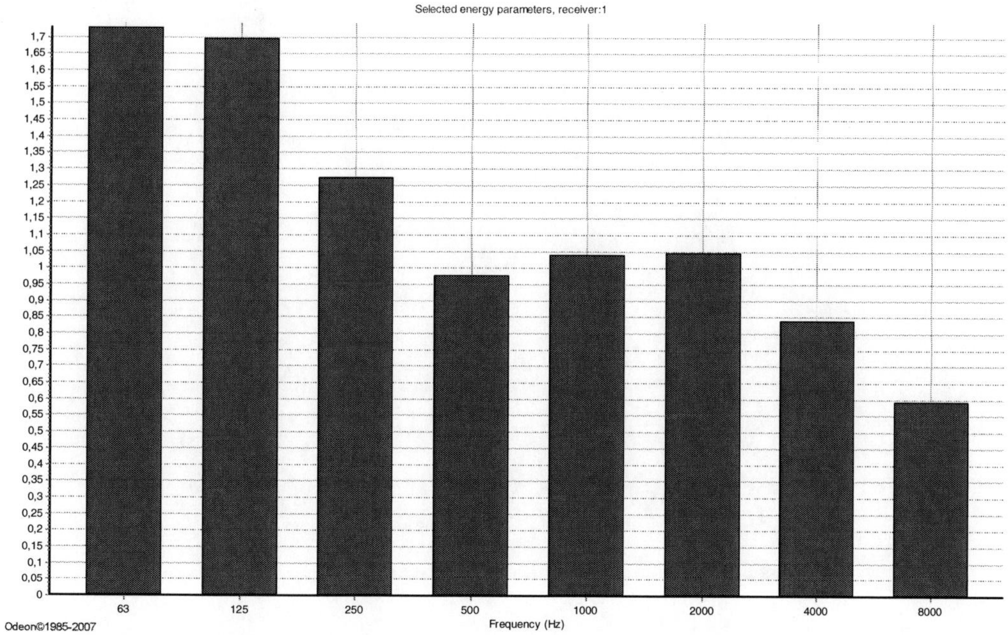

Figure 6. Simulated RT for fitness gym 02 with acoustically treated ceiling.

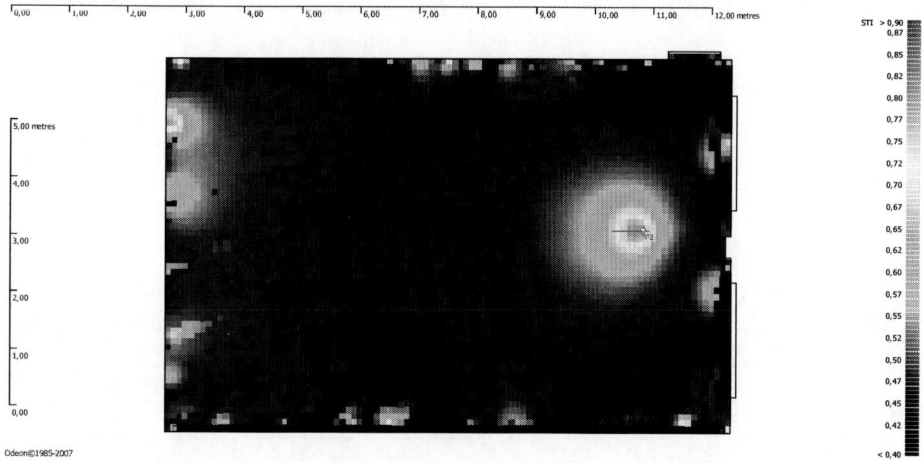

Figure 7. Simulated STI for fitness gym 02 with original ceiling.

Table 6. Sound levels measured in fitness gyms

	L_{min} dB(A)	L_{eq} dB(A)	L_{max} dB(A)
Fitness gym 01	58	82	107.9
Fitness gym 02	60.2	85.5	108.8
Fitness gym 03	58.5	74.3	92.1
Fitness gym 04	56.6	80.1	96.4

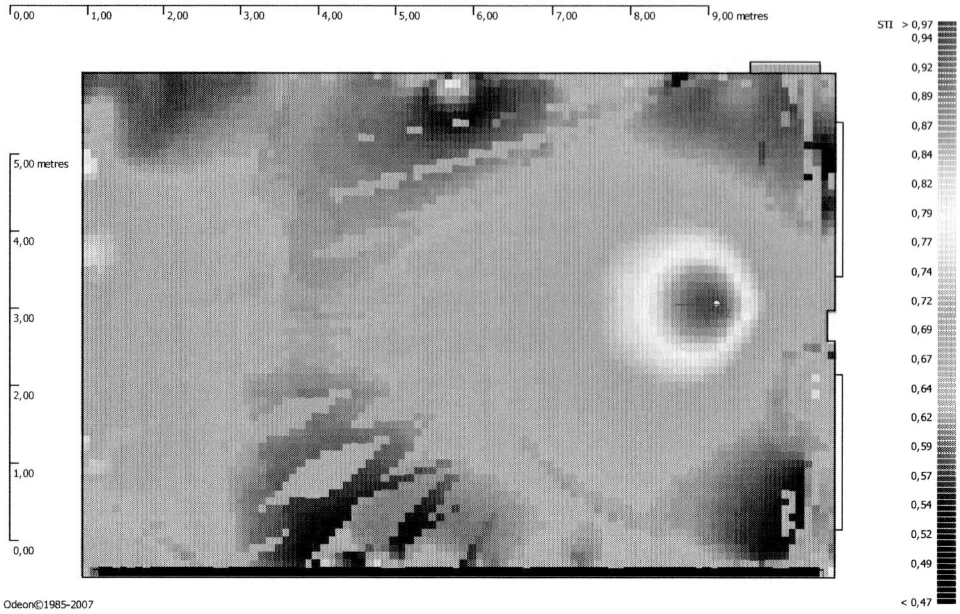

Figure 8. Simulated STI for fitness gym 02 with modified ceiling.

Table 7. Limits of sound levels in the evaluation of occupational noise in different countries (adapted from Johnson et al., cited by Goelzer et al., 2001)

Countries	Daily exposure time 8-hour average sound pressure level dB(A)	Daily exposure time 8-hour average maximum sound pressure level for auditory monitoring dB(A)
Brazil	85	85
Netherlands	85	80
Norway	85	80
Spain	85	80
Sweden	85	80

Figures 5, 6, 7 and 8 show the RT and STI in simulated fitness gym 02 with the original ceiling and the acoustically treated ceiling.

Table 6 lists the sound levels measured in the four fitness gyms, while Table 7 describes the recommended sound levels limits in different countries for the conditions of daily occupational exposure and for auditory monitoring.

The equivalent sound levels measured in fitness gyms 01, 02 and 04 were high, reaching close to or exceeding the occupational exposure limit of 85 dB(A), as shown in Table 7.

Table 6 also shows very high values of maximum measured noise levels, especially in fitness gyms 01 and 02, where the Lmax reached values of 107.9 dB(A) and 108.8 dB(A), respectively. According to the Brazilian Regulatory Standard NR 15 (2011), which evaluates noise levels in workplace environments (e.g., industrial environments), the maximum permissible exposure time to sound levels of around 108 dB(A) is approximately 20 minutes.

Since personal trainers (see data below) usually work an average of 4:25 hours per day in fitness gyms, occupational insalubrity is evident (Andrade and Russo, 2010; NR15, 2011).

Table 7 shows that many countries consider a noise level of 80 dB(A) the limit for beginning auditory monitoring in workplace environments. An analysis of Table 6 indicates that, with the exception of fitness gym 03, all the others showed noise levels higher than or equal to 80 dB(A). The paradox here is that environments intended to improve the quality of life and health of their patrons show sound levels which should be monitored to protect their hearing, i.e., the occupants of these environments – both personal trainers and users – should be wearing hearing protectors.

The effects of noise on gym users and personal trainers were investigated based on a sample comprising 18 individuals, 04 personal trainers and 14 users, all of them women. The ages of the personal trainers ranged from 26 to 43 years, with an average age of 30.5. The ages of the users ranged from 19 to 67, with a mean age of 30.6. As for the time the users spent in the fitness gyms, the average time was 3:09 hours, although this number varied in each fitness gym, as indicated in Table 8.

While performing physical activities accompanied by loud music, the users interviewed reported experiencing headaches, dizziness and vocal fatigue. On the other hand, the complaints most frequently reported by users after completing their physical exercises were vocal fatigue, stomach pain, and dizziness. These complaints were also observed in a study by Andrade et al. (2009).

With regard to the personal trainers, when asked how long they had worked at the fitness gyms under evaluation, the reported time ranged from 1 month to 15 years. The time spent per week in exercise rooms varied from 02 to 06 hours, with an average time of 4:25 hours.

As for the effects of loud music during the performance of physical exercise, it was found that 25% of the personal trainers reported vocal fatigue (see Figure 11). However, contrary to the findings of studies such as those of Deus and Duarte (1997), Andrade and Russo (2010) and Martines and Bernardi (2001), the personal trainers did not report feeling headaches during their work activities.

After completing the physical exercises, the complaints of vocal fatigue increased from 25% to 50% (see Figure 12). This may be due to the considerable effort the physical trainer must make to raise her voice, since it has to compete with loud music (see Table 6).

Several studies associate the activities of physical education professionals with complaints related to vocal alterations (Heidel and Torgenson, 1993; Long et al., 1998; Smith et al. 1998; Stefanni et al., 2011). In the study by Farias and Noel (2004), 100% of the interviewees reported severe discomfort at the end of the workday, presenting symptoms such as phlegm, dry cough, burning, hoarseness, vocal fatigue and increased effort to speak. Zucki et al. (2006) argue that 63.3% of personal trainers report voice problems in their profession. Sas et al. (2007) stated that 70.8% of personal trainers complained of vocal fatigue and 58.3% reported hoarseness at the end of the workday.

According to Silva et al. (2009) and Palma et al. (2009), it is recommended that personal trainers use a microphone to minimize the vocal effort expended in gym classes.

Table 8. Average number of hours the user spends in the gym room

Fitness gym 01	4:06
Fitness gym 02	2:25
Fitness gym 03	3:05
Fitness gym 04	3:00

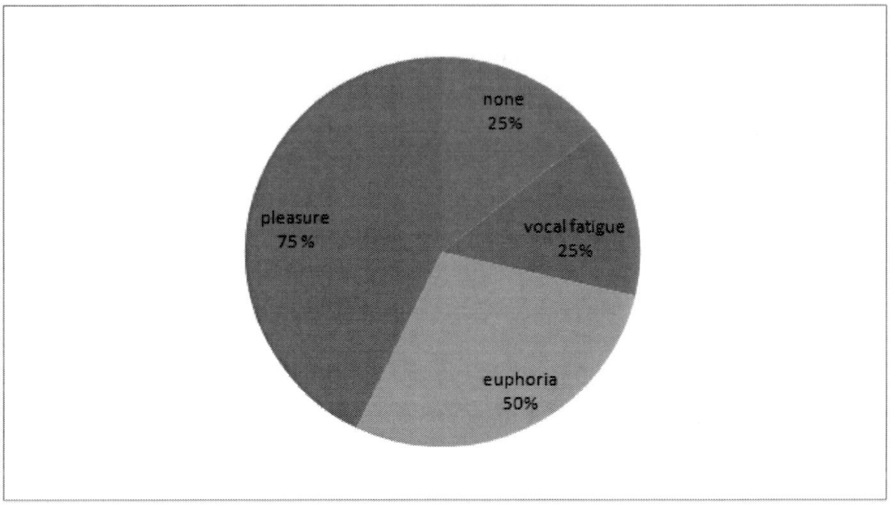

Figure 11. Effects of noise on personal trainers during physical activity.

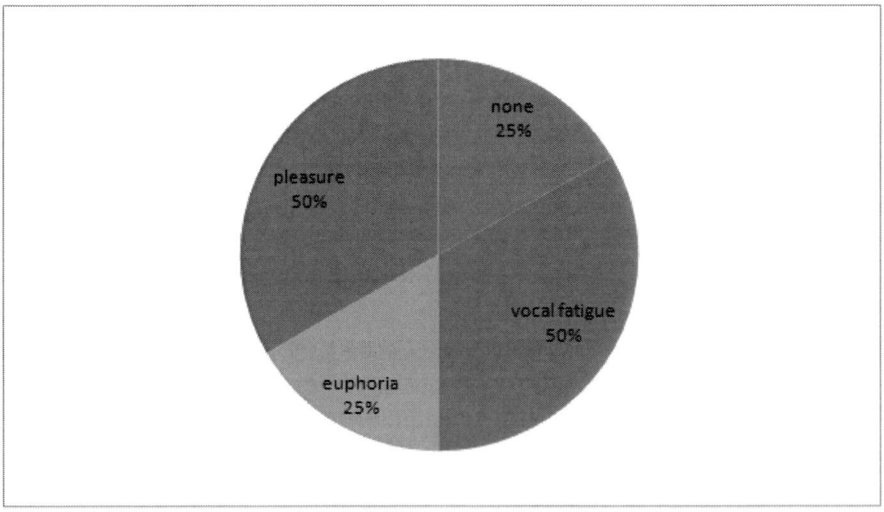

Figure 12. Effects of noise on personal trainers after physical activity.

CONCLUSION

This study shows that environments where people seek health and leisure present high noise levels, which can be classified as a cause for extreme concern with regard to the health of their patrons, and especially that of the professionals who work there. From the acoustic standpoint, fitness gyms can be characterized as unhealthy environments for both users and personal trainers, since measured sound pressure levels reached the maximum level of Lmax 108.8 dB(A), which is a noise level usually found in industrial environments.

The results of this study indicate the need for noise control through various measures such as abatement of the sound levels of loud music (noise control at source), and acoustic treatment of gyms (control of sound trajectories).

Computer simulation offers a reliable way to determine suitable acoustic conditions for workplace environments. In the case in question, substituting the ceiling finishing material for another with a higher absorption coefficient reduced the reverberation time, and speech transmission index, changing their classification from Poor to Good.

The improvement in the conditions of intelligibility in the work environment under study or in similar environments may benefit users and personal trainers, for example, by diminishing their vocal fatigue. Vocal fatigue was pinpointed by personal trainers and users as one of the main deleterious effects resulting from noise in fitness gyms.

Given these findings and the current literature, it is suggested that personal trainers working in fitness gyms use microphones to minimize voice stress during classes. In addition, it is recommended to consult an audiologist, who should consider auditory monitoring by means of initial and periodic hearing tests, as well as correct voice preparation for the performance of work activities in fitness gyms.

ACKNOWLEDGMENTS

The authors gratefully acknowledge the German Government, through the German Academic Exchange Service – DAAD (Deutscher Akademischer Austauschdienst) and the Brazilian Government, through the National Council for Scientific and Technological Development – CNPq, for their financial support, which enabled the purchase of the sound level meters and software used in this study.

REFERENCES

Alberta Infrastructure and Transportation – Standards and Guidelines for School Facilities, 2007, 22-24.

Ananthaganeshan KA, Gastmeier WJ. Acoustical performance criteria, treatment and guidelines for multifunctional school gymnasia. *Canadian Acoustics,* 2007, 35(4): 25-30.

Andrade IFC, Russo ICP. Relação entre os achados audiométricos e as queixas auditivas e extra-auditivas dos professores de uma academia de ginástica. [Relationship between audiometric findings and auditory and extra-auditory complaints of physical educators in a fitness gym] *Revista da Sociedade Brasileira de Fonoaudiologia,* 2010, 15, 167-173 (in Portuguese).

Andrade IFC, Souza AS, Frota SMMC. Estudo das emissões otoacústicas – produto de distorção durante a prática esportiva associada à exposição à música. [Otoacoustic emissions – a product of distortion during sports activities associated with exposure to music] *Revista CEFAC,* 2009, 11(4): 654-661 (in Portuguese).

Associação Brasileira de Normas Técnicas. NBR 10151: Avaliação do ruído em áreas habitadas visando o conforto da comunidade – procedimento. [Brazilian Association of Technical Standards. NBR 10151: Noise assessment in populated areas to ensure the comfort of the community – Procedure] Rio de Janeiro, 2000 (in Portuguese).

Brattico E, Kujala T, Tervaniemi M, Alku P, Ambrosi L, Monitillo V. Long-term exposure to occupational noise alters the cortical organization of sound processing. *Clin. Neurophysiol.* 2005, 116:190-203.

Chiovenda P, Pasqualetti P, Zappasodi F, Ercolani M, Milazzo D, Tomei G, Capozzella A, Tomei F, Rossini PM, Tecchio F. Environmental noise-exposed workers: event-related potentials, neuropsychological and mood assessment. *International Journal of Psychophysiology*, 2007, 65:228-237.

Comitê Nacional de Ruído e Conservação Auditiva. Boletim nº 1. [National Committee on Noise and Hearing Conservation. Bulletin No. 1] São Paulo, 1994 (in Portuguese).

Costa EA, Morata TC, Kitamura S. Patologia do ouvido relacionada com o trabalho. [Work-related hearing pathology] In: Patologia do Trabalho. Mendes R. [Ed].São Paulo, Atheneu, 2003 (in Portuguese).

Deus MJ, Duarte MS. Nível de pressão sonora em academias de ginástica e a percepção auditiva dos professores. [Sound pressure level in fitness gyms and auditory perception of personal trainers] *Revista Brasileira de Atividade Física e Saúde,* 1997, 2 (2): 05-16 (in Portuguese).

Farias E, Noel KO. Análise preliminar da saúde vocal de profissionais de Educação Física como fator inerente à sua atuação em academias. [Preliminary analysis of the vocal health of Physical Education professionals as a factor inherent to their performance in fitness gyms] 2004. Available at: http://www.edvaldodefarias.com/saudevocal.pdf. Viewed on 25 May 2012 (in Portuguese).

Fiorini AC. Impacto Ambiental e Ocupacional na Audição de Trabalhadores. [Environmental and Occupational Impact on Workers' Hearing] In: Fernandes FDM et al. [Ed]. Tratado de Fonoaudiologia. [Treatise on Phonoaudiology] São Paulo: Editora Roca, 2009 (in Portuguese).

Gonçalves CGO. Saúde do Trabalhador – da estruturação à avaliação de programas de preservação auditiva. [Worker Health – from the structuring to the evaluation of hearing preservation programs] São Paulo: Editora Roca, 2009 (in Portuguese).

Harris CM. Handbook of acoustical measurements and noise control. USA: Acoustical Society of America, 1998.

Harris CM. Noise Control in Buildings. McGraw-Hill, Inc. New York, USA, 1994.

Heidel SE, Torgerson JK. Vocal problems among aerobic instructors and participants. *Journal of Communication Disorders*, 1993, 26(3):179-191.

International Electrotechnical Commission. IEC 60268- 16: Sound system equipment- Part 16: Objective rating of speech intelligibility by speech transmission index. Switzerland, 2003.

International Organization for Standardization. ISO 3382-1: Acoustics — Measurement of room acoustic parameters - Part 1: Performance spaces. Switzerland, 2009.

International Organization for Standardization. ISO 3382-2: Acoustics-Measurement of room acoustic parameters - Part 2: Reverberation time in ordinary rooms. Switzerland, 2008.

Isover do Brasil – Catálago de produto do fabricante [Manufacturer's product catalogue]. www.isover.com.br (in Portuguese).

Johnson DL, Papadopoulos P, Watfa N, Takala J. Exposure Criteria, Occupational Exposure Levels. In: Occupational Exposure to Noise: Evaluation, Prevention and Control. [Ed] Goelzer B, Hansen CH, Sehrndt, GA. Berlin, WHO, 2001.

Kang J. Numerical modeling of speech intelligibility in dining spaces. *Applied Acoustics,* 2002, 63: 1315-1333.

Karabiber K, Vallet M. Classroom acoustics policies – an overview. In: Euronoise Naples, Italy, 2003.

Kós AOA, Kós MI. Etiologias das Perdas Auditivas e suas Características Audiológicas. [Etiology of Hearing Loss and its Audiological Characteristics] In: Frota, S. Fundamentos em Fonoaudiologia – Audiologia. [Fundamentals of Phonoaudiology - Audiology] Rio de Janeiro: Editora Guanabara Koogan, 1998 (in Portuguese).

Kujala T, Yuri S, Winkler I, Saher M, Tervaniemi M, Sallinen M, Teder-Sälejärvi W, Alho K, Reinikainen K, Näätänen R. Long-term exposure to noise impairs cortical sound processing and attention control. *Psychophysiology,* 2004; 41: 875-881.

Lacerda ABM. Caracterização dos níveis de pressão sonora em academias de ginástica e avaliação auditiva de seus professores. [Characterization of sound pressure levels in fitness gyms and auditory evaluation of their personal trainers] Universidade Tuiuti do Paraná, 1999. Master's Dissertation (in Portuguese).

Long J, Williford HN, Olson MS, Wolfe V. Voice problems and risk factors among aerobics instructors. *Journal of Voice*, 1998, 12(2):197-207.

Maia AA, Gonçalves DU, Menezes LM, Barbosa BMF, Almeida OS, Resendo LM. Análise do perfil audiológico dos músicos da Orquestra Sinfônica de Minas Gerais (OSMG). [Analysis of the audiological profile of the musicians of the of Minas Gerais Philharmonic Orchestra (OSMG)] Per Musi – Revista Acadêmica de Música, Belo Horizonte, 2007, 15:67-71 (in Portuguese).

Marcon CR, Zannin PHT. Avaliação do ruído gerado por academias de ginástica. [Evaluation of noise generated in fitness gyms] *Engenharia e Construção, Curitiba,* 2004, 96: 39-42 (in Portuguese).

Martines CR, Bernardi APA. A percepção diferenciada do barulho: estudo comparativo com jovens frequentadores e funcionários de casas noturnas da cidade de São Paulo. [A different perception of the noise: a comparative study with young employees and patrons of nightclubs in São Paulo] *Revista CEFAC*, 2001 3:71-76. in Portuguese).

Matos ECG. Incômodo decorrente do ruído na escola em professores com e sem distúrbios vocais. [Discomfort caused by noise in school for teachers with and without voice disorders] Pontifícia Universidade Católica do Paraná, 2008. Master's Dissertation (in Portuguese).

MelnickW. Saúde auditiva do trabalhador. [Worker hearing health] In: Katz, J. Tratado de Audiologia Clínica. [Treatise on Clinical Audiology] São Paulo: Editora Manole, 1999 (in Portuguese).

NR – 15 - Norma Regulamentadora NR-15 – Atividades e operações insalubres, Anexo 1 [Regulatory Standard NR-15 – Unhealthy activities and operations, Appendix 1]. Segurança e Medicina do Trabalho [Safety and Occupational Medicine], Editora Atlas, São Paulo; 68th ed., 2011 (in Portuguese).

Nudelmann, AA et al. [Ed]. PAIR – Perda auditiva induzida pelo ruído. [NIHL – Noise-induced hearing loss] Porto Alegre: Bagagem Comunicação, 1997 (in Portuguese).

ODEON Software Handbook - Version 9.2 - Combined.

Packer K, Faszer C. Reverberation measurement and prediction in gymnasia with non-uniformly distributed absorption; The importance of diffusion. *Canadian Acoustics/Acoustique canadienne*, 2011, 39 (1): 31-37.

Palma A, Mattos UA, Almeida MN, Oliveira GEMC. Nível de ruído no ambiente de trabalho do professor de educação física em aulas de ciclismo indoor. [Noise level in the work environment of the physical education teacher in indoor cycling classes] *Revista de Saúde Pública*, 2009, (43), 2: 345-351 (in Portuguese).

Passero C, Zannin PHT. Statistical comparison of reverberation times measured by the integrated impulse response and interrupted noise methods, computationally simulated with ODEON software, and calculated by Sabine, Eyring and Arau-Puchades formulas. *Applied Acoustics,* 2010, 71, 1204-1210.

Penney PJ, Earl C. Occupational noise and effects on blood pressure: exploring the relationship of hypertension and noise exposure in workers. *AAOHN J,* 2004; 52(11): 476-480.

Pinto, PM, Russo, ICP. Estudo dos efeitos da exposição à música excessivamente amplificada sobre a audição de professores de academias de ginástica. [Study of the effects of exposure to excessively loud music on the hearing of personal trainers in fitness gyms] *Revista CEFAC*, 2001, 3: 65-69 (in Portuguese).

Ronsse L. Speech intelligibility of lecture halls. *Journal of the Acoustical Society of America,* 2006, 119 (5): 3207.

Russo, ICP. Acústica e Psicoacústica aplicadas à Fonoaudiologia. [Acoustics and Psychoacoustics applied to Phonoaudiology] São Paulo: Editora Lovise, 1999 (in Portuguese).

Russo, ICP. Noções gerais de acústica e psicoacústica. [General notions of acoustics and psychoacoustics] In: Nudelmann, AA. et al. [Ed] PAIR – Perda auditiva induzida pelo ruído. [NIHL – Noise-induced hearing loss] Porto Alegre: Bagagem Comunicação, 1997 (in Portuguese).

Russo ICP. Noções gerais de acústica e psicoacústica. [General notions of acoustics and psychoacoustics] In: Nudelmann, AA. et al. [Ed] PAIR – Perda auditiva induzida pelo ruído. [NIHL – Noise-induced hearing loss] Porto Alegre: Bagagem Comunicação, 1997 (in Portuguese).

Sas RM, Fenske SG, Dugnani K, Pizzani L, Almeida MA, Costa MPR. Professores de Educação Física: Incidência de alterações vocais. [Physical Education Teachers: Incidence of vocal alterations] Revista Uniube, 2007. Available at: http://revistas.uniube.br/index.php/anais/article/view/332. Viewed on 14 Mar 2012 (in Portuguese).

Silva PSB, Ferreira CES, Cavalcante MM, Garavelli, SL. Nível de ruído sonoro nas aulas de ciclismo indoor em academias do Distrito Federal. [Sound level of noise in indoor bike classes at gyms Federal District] Educação Física em Revista, 2009. Available at:

http://www.portalrevistas.ucb.br/index.php/efr/issue/view/119. Viewed on 17 Apr 2012 (in Portuguese).

Smith E, Kirchner HL, Taylor M, Hoffman H, Lemke JH. Voice problems among teachers: differences by gender and teaching characteristics. *Journal of Voice,* 1998, 12(3):328-334.

Steffani JA, Vieceli,VCB, Grasel, CE. Saúde vocal e nível de atividade física dos professores de Educação Física. [Vocal health and level of physical activity of Physical Education teachers] Revista Digital Buenos Aires, 2001. Available at: http://www.efdeportes.com/efd155/saude-vocal-dos-profissionais-de-educacao-fisica.htm. Viewed on 23 Apr 2012 (in Portuguese).

World Health Organization – WHO - Résumé d'Orientation des Directives de l'OMS Relatives au Bruit dans l'Environmental, 2003.

World Health Organization. Burden of disease from environmental noise. Copenhagen, 2011.

Yonzaki C, Hidaka UM. Inserção da Fonoaudiologia na saúde do trabalhador. [Inclusion of Phonoaudiology in worker health] In: Lopes, O. Tratado de Fonoaudiologia. [Treatise on Phonoaudiology] 2nd ed. São Paulo: Editora Tecmed, 2005 (in Portuguese).

Zannin PHT, Passero CRM, Sant'Ana DQ. Acoustic Design of Enclosed Spaces. [in] Ergonomics: Design, Integration and Implementation. Bram N. Brinkerhoff [Ed]. New York: Nova Science Publishers, 2009.

Zannin PHT, Passero CRM, Sant'Ana DQ., Bunn F, Fiedler PEK, Ferreira AMC. Classrooms Acoustics: Measurements, Simulations and Applications. [in] Classrooms: Management, Effectiveness and Challenges. Rebecca J. Newley [Ed]. New York: Nova Science Publishers, 2011.

Zannin PHT, Zwirtes DP, Passero CRM. Assessment of Acoustic Quality in Classrooms - Based on Measurements, Perception and Noise Control. [in] Noise Control, Reduction and Cancellation – Solutions in Engineering. Daniela Siano [Ed]. Croatica: InTech – Open Science. www.intechopen.com, 2012.

Zucki F, Lacerda ABM. O ruído em atividades de educação física. [The noise in physical education activities] In: Morata TC, Zucki F. [Ed] Saúde Auditiva – Avaliação de Riscos e Prevenção. [Hearing Health - Risk Assessment and Prevention] São Paulo: Plexus, 2010 (in Portuguese).

Zucki, F; Morata, T, Mendes. Percepção de estudantes, profissionais e coordenadores de graduação de educação física sobre o ruído em sua profissão. [Perception of students, professionals and coordinators of undergraduate physical education on noise in their profession] *Revista da Sociedade Brasileira de Fonoaudiologia,* 2006 (11)4:250-264 (in Portuguese).

In: Noise and Ergonomics in the Workplace
Editor: Paulo Henrique Trombetta Zannin

ISBN: 978-1-62417-573-2

Chapter 10

EVALUATION OF SOUND PRESSURE LEVELS IN DIGITAL AUDIO PLAYERS

Paulo Henrique Trombetta Zannin*, Flávio Delarte Levis and Rodrigo Raphael Raffler

Laboratory of Environmental and Industrial Acoustics and Acoustic Comfort (LAAICA), Federal University of Paraná, Brazil

ABSTRACT

An evaluation was made of the sound levels generated by digital players selected among the models available in the Brazilian market. Measurements were taken of the equivalent sound pressure level L_{eq}, the maximum sound pressure level L_{Max} and the minimum sound pressure level L_{Min}, all expressed in dB(A). The results indicate that the noise levels generated by these players exceed the levels considered safe for human health. They also show that in one hour of use of digital players it is possible to exceed the noise dose of 100% that is expected to occur in an 8-hour workday, according to occupational health evaluation standards.

INTRODUCTION

Urban growth and the ever increasing number of sound sources (buses, cars, construction sites, planes, trains, neighbors, leisure activities, etc.) have all contributed to increase environmental noise levels (Griffiths, I.D. and Langdon, F.J., 1986; Filho JMA et al., 2004; da Paz E et al., 2005; Lebiedowska, B., 2005; Zannin PHT et al., 2006; da Paz E and Zannin PHT, 2010). Among these sources, leisure involving sports is an ever present source of sound in urban environments and is intrinsically linked to human activities (Zannin PHT et al., 2002; Szeremeta B and Zannin PHT, 2009). Within this context we find that music, a cultural

* paulo.zannin@pesquisador.cnpq.br; paulo.zannin@gmail.com.

element that has accompanied mankind's evolution, is an integral part of the most diverse societies.

A recent issue of *Época* magazine (2009) published a report entitled "*Era of the digital deaf*", alluding to the indiscriminate use of audio players at sound volumes that phonoaudiologists consider extremely harmful to auditory health. The problem caused by the uncontrolled use of these devices has led to the so-called "*iPod disorder*", which is already well known in countries such as the U.S., China, Mexico and Brazil, among others (Revista Época, 2009).

The daily newspaper *Jornal do Brasil* (2009) also published an article in October 2009 about the problem resulting from the use of digital audio players (DAPs). According to this newspaper, researchers in the European Union estimate that 2.5 to 10 million Europeans may have sustained hearing impairment from listening to DAPs that emit sound levels exceeding 89 dB(A). This prediction by European specialists presupposes the use of MPs players for more than one hour a day for a period of at least five years (*Jornal do Brasil*, 2009).

In this context, this study reports on measurements of sound levels generated by two digital players available on the Brazilian market. For reasons of confidentiality, the names of the manufacturers are not mentioned here, and the devices are referred to as Player A and Player B.

MATERIALS AND METHODS

Two models available on the Brazilian market were chosen to evaluate the sound levels generated by the DAPs. These models are referred to herein as Player A and Player B (see Figure 1).

In market terms, Player A is considered an original product, while Player B is considered a "*generic*" version of Player A. The earphone used with Player B is the same brand as the one used with Player A and was also used for the measurements of Player B.

Figure 1. Left – Player A and Right – Player B.

The sound levels generated by the DAPs were measured with a Brüel and Kjaer 2238 sound level meter, which recorded the equivalent sound pressure level L_{eq}, the maximum sound pressure level L_{Max} and the minimum sound pressure level L_{Min}, all expressed in dB(A). The background noise was measured for 10 minutes at the Acoustics Laboratory of the Federal University of Paraná prior to beginning the evaluation of the DAPs.

The sound levels of the players were measured on a volume scale of 3 to 30, in intervals of 3, as indicated in Table 1. A piece of the same music was selected and played for 1 minute on each of the players. The volume of the player was gradually modified on a scale of 3 points up to a maximum of 30 (see Table 1).

The noise dose was calculated using the equation below, according to the Occupational Health Standard of *Fundacentro* (NHO-01, 1999):

$$D = \frac{T_E}{480} \cdot 100 \cdot 2^{\left(\frac{NE-CR}{q}\right)} \quad [\%] \tag{1}$$

Where: D is the daily noise dose expressed in percent [%]; T_E = exposure time during a workday [min]; CR is the reference criterion expressed in dB(A); NE is the noise level of exposure in dB(A), and q is the dose duplication increment.

In Brazil, according to Regulation Act NR 15 of the Ministry of Labor and Employment (Safety and Occupational Medicine, 2011), the criterion of reference is de CR = 85 dB(A), and this standard uses the dose duplication factor of q = 5. Fundacentro's Occupational Health Standard (NHO-01, 1999) uses q = 3 and the reference criterion CR = 85 dB(A). The ISO 1999 standard (1990) – *Acoustics – Determination of occupational noise exposure and estimation of noise-induced hearing impairment* – also uses q = 3.

Starting from these data, the noise dose was calculated based on an exposure time of 1 hour. This calculation considered the dose duplication factor of q = 3, and the reference criterion CR = 80 dB(A). This reference criterion is adopted in countries such as Spain, the Netherlands, Norway and Sweden (Johnson et. al., 2001). It refers to the ultimate sound pressure level that should be observed and respected in auditory monitoring for occupational health, considering an exposure time of 8 hours. This reference criterion is stricter than the CR = 85 dB(A), and is therefore used here, since what is being considered is the auditory health of millions of people, especially children and youths, who use digital audio players but are unaware of the possible harmful effects on their overall health and particularly on their auditory health (*Revista Época*, 2009; *Jornal do Brasil*, 2009).

Results and Discussion

The measurement of background noise at the Federal University of Paraná's Acoustics Laboratory indicated a value of L_{eq} 37.3 = dB(A) for a measured time of 10 minutes and was therefore considered a suitable environment for the measurements, since the background noise level was well below the sound levels generated by the digital audio players.

The sound levels generated by the players were measured using the above described procedure. Table 1 lists the results of these measurements.

Table 1. Sound levels measured in the digital audio Players A and B

Player A				Player B			
Volume	L_{eq} dB(A)	L_{Max} dB(A)	L_{Min} dB(A)	Volume	L_{eq} dB(A)	L_{Max} dB(A)	L_{Min} dB(A)
3	43.8	63.2	33.4	3	47.3	55.3	35.7
6	55.7	64.2	39.4	6	51.3	59.7	35.1
9	62.6	71.2	36.6	9	56.3	64.7	35
12	68.6	77.2	51.3	12	60.9	69.3	35.6
15	74.7	83.1	57.2	15	66.3	74.4	33.1
18	80.6	89.1	57.5	18	71.4	79.8	36.3
21	84.5	93	37.6	21	75.7	83.9	35.7
24	87.4	96	34.6	24	80.1	88.3	35.9
27	90.1	99	34.8	27	85.4	93.6	67.6
30	91.2	100.8	57.6	30	90.5	98.8	34.7

Table 1 demonstrates how high the sound levels generated by Players A and B are, especially starting from Volume = 12 in Player A and Volume 15 in Player B. Belojevic et al. (1997) report that individuals exposed to noise levels exceeding L_{eq} = 65 dB(A) are in a high-risk zone where the negative influences of noise lead to sleep disorders, psychological disorders and behavioral disorders. Época magazine (2009) published a report about an adolescent who claimed to have slept an entire night with his MP3 player turned on in his ears. Farina (2007) has evaluated 13 models of DAPs available in the italian Market, and used by youngsters of 15-18 years of age, and concluded that there is indeed a real risk of hearing health damage promoted by several of the evaluated DAPs.

The equivalent sound pressure levels L_{eq} listed in Table 1 were used to calculate the noise dose delivered in 1 hour of use of the player. Using equation (1), the calculations indicated that Player A operating at Volume = 27 and Volume = 30 delivered a noise dose level of 127.9% and 164.8%, respectively. The limit noise dose of 100% recommended by the NR-15 and NHO-01 standards for work environments, taking into account 8 hours of exposure per day, was exceeded in one hour of exposure to Player A. In the case of Player B, the limit dose of 100% was exceeded at Volume = 30, reaching a value of 140.3% (see Table 2).

Table 2. Noise dose calculated for 1 hour of use of the digital audio players

Volume	Player "A" Noise dose (%)	Player "B" Noise dose (%)
3	0.003	0.0067
6	0.046	0.017
9	0.23	0.053
12	0.91	0.15
15	3.69	0.53
18	14.4	1.73
21	35.2	4.64
24	68.7	12.8
27	*127.9*	43.3
30	*164.8*	*140.3*

Table 3. Exposure time T_E required to obtain the dose limit - D= 100%

Volume	Limit of Exposure Time T_E (min)	
	Player A	*Player B*
3	2,059,064	917,209
6	131,700	363,995
9	26,743	114,651
12	6,686	39,609
15	1,633	1,1375
18	418	3,501
21	170	1,296
24	87	469
27	47	138
30	36	42

Table 3 shows the exposure time that would be required to reach the 100% limit dose for 8 daily hours of exposure, according to the rules of occupational noise standard NHO-01, at the volumes of the DAPs under analysis. As can be seen, as the noise level is raised from 3 to 30, the allowed time of exposure decreases. Approximately 47 minutes would be required with Player A to reach the 100% dose at Volume = 27, whereas approximately 36 minutes would be required at Volume = 30.

Although the noise levels listed in Table 1 are high, these levels can be and are reached every day. Época magazine (2009) describes some of the reasons that may lead users to listen to their players at high sound levels: 1) The external noise level is very high, so the player's sound level is increased to mask it, 2) A desire to isolate oneself from the outside world, 3) The need to take full advantage of the player's acoustic power level, 4) Listening to music at low sound levels is not stimulating. These are the cases which pose health risks to users, who do not realize or even know that they are harming their hearing. In this context, the very least the manufacturers of these devices should do is to warn users of their potentially harmful effects on hearing, and to establish safety limits for the sound levels to be used, as well as for the exposure time to these devices.

CONCLUSION

The sound pressure levels measured in the two digital audio players evaluated in this study demonstrate that, starting from volume 12 in Player A and from volume 15 in Player B, the sound levels already exceed a L_{eq} = 65 dB(A). This level of sound pressure is used by the Brazilian Standard on Ergonomics NR 17 (Ergonomics, Safety and Occupational Medicine, 2011) to evaluate whether a working place is comfortable or not. Noise levels above 65 dB(A) render an environment uncomfortable. Measurements also indicate that the noise levels generated by these players exceed the levels considered safe for human health. They also show that in one hour of use of digital players it is possible to exceed the noise dose of 100%

that is expected to occur in an 8-hour workday, according to occupational health evaluation standards.

It is imperative that awareness be raised concerning the possible impairment of the auditory health of users of digital audio players. In this sense, the action of public health agencies in publicizing the existence of this problem is also necessary. As for the manufacturers of these devices, they should adopt measures to limit the use of DAPs at high sound levels, and should also inform users about possible noise-induced hearing impairment if the ultimate sound pressure level is exceeded.

Acknowledgment

The authors gratefully acknowledge the German Government, through the German Academic Exchange Service – DAAD (Deutscher Akademischer Austauschdienst) and the Brazilian Government, through the National Council for Scientific and Technological Development – CNPq, for their financial support, which enabled the purchase of the sound level meters and software used in this study.

References

Belojevic G, Jakovljevic B, Aleksic O. Subjective reactions to traffic noise with regard to some personality traits. *Environment International* 1997, 23:221-226.

da Paz E, Ferreira AMC, Zannin PHT. Comparative study of the perception of urban noise (Estudo comparativo da percepção do ruído). *Revista de Saúde Pública* 2005, 39: 467-472 (in Portuguese).

da Paz E, Zannin PHT. Urban daytime traffic noise prediction models. *Environmental Monitoring and Assessment* 2010, 163: 515-529.

Filho JMA, Lenzi A, Zannin PHT. Effects of traffic composition on road noise: A case study. *Transportation Research Part D: Transport and Environment* 2004, 9: 75-80.

Fundacentro – Fundação Jorge Duprat Figueiredo de Segurança e Medicina do Trabalho - Norma de Higiene Ocupacional. Procedimento Técnico. *Avaliação da exposição ocupacional ao ruído*, 1999. (in Portuguese)

Farina A. A study of hearing damage caused by personal MP3 players. Audio Engineering Society - Convention Paper. Presented at the 123rd Convention 2007 October 5–8 New York, NY, USA.

Johnson DL, Papadopoulos P, Watfa N, Takala J. Exposure Criteria, Occupational Exposure Levels. In: Occupational Exposure to Noise: Evaluation, Prevention and Control. [Ed] Goelzer B, Hansen CH, Sehrndt GA. Special Report S 64 - World Health Organization, WHO, Berlin, 2001.

Griffiths I.D., Langdon F.J. Subjective response to road traffic noise, *Journal of Sound and Vibration* 1986, 16: 32-8.

International Organization for Standardization - ISO 1999 - Acoustics – Determination of occupational noise exposure and estimation of noise-induced hearing impairment, Geneve 1990.

Jornal do Brasil – JB Vida, Saúde e Ciência – "Audição – União Européia quer diminuir volume de players", Ano 119 - Nº 117, Rio de Janeiro 12 de outubro de 2009. (in Portuguese)

Lebiedowska B. Acoustics background and transportation noise in urbanized areas: A note on the relative classification, *Transportation Research Part D - Transport and Environment* 2005, 341: 345-10.

Norma Regulamentadora NR-15 – Atividades e operações insalubres, Anexo 1 [*Regulatory Standard NR-15 – Unhealthy activities and operations, Appendix 1*]. Segurança e Medicina do Trabalho [*Safety and Occupational Medicine*], Editora Atlas, São Paulo; 68th ed., 2011 (in Portuguese)

NR -17 - Norma Regulamentadora NR-17 – Ergonomia, Segurança e Medicina do Trabalho [*Regulatory Standard NR-17 – Ergonomics, Safety and Occupational Medicine*], Editora Atlas, São Paulo; 68th ed., 2011 (in Portuguese)

NHO-01 – Fundacentro - NHO-01: Norma de Higiene Ocupacional, avaliação da exposição ocupacional ao ruído [*Occupational Hygiene Standard, assessment of occupational noise exposure*]. São Paulo, 1999. (in Portuguese).

Revista Época – "A era dos surdos digitais" – 06 de abril de 2009, Nº 568, Editora Globo. (in Portuguese)

Szeremeta B, Zannin PHT. Analysis and Evaluation of Soundscapes in public parks through interviews and measurement of noise. *Science of the Total Environment* 2009, 407: 6143-6149.

Zannin PHT, Calixto A, Diniz FB, Ferreira JAC, Schuhli RB. Annoyance caused by urban noise to the citizens of Curitiba, Brazil (Incômodo causado pelo ruído urbano a população de Curitiba, Pr). *Revista de Saúde Publica* 2002, 36: 521-524. (in Portuguese).

Zannin PHT, Ferreira AMC, Szeremeta B. Evaluation of noise pollution in urban parks. *Environmental Monitoring and Assessment* 2006, 118: 423-433.

In: Noise and Ergonomics in the Workplace
Editor: Paulo Henrique Trombetta Zannin

ISBN: 978-1-62417-573-2

Chapter 11

Assessment of Occupational Exposure to Whole-Body Vibration: Application of ISO Standards 2631-1 and ISO 2631-5

***Bruno Sérgio Portela*[1,2*] *and Paulo Henrique Trombetta Zannin*[2]**

[1]Laboratory of Biomechanics and Ergonomics,
Midwestern State University of Paraná, Brazil
[2]Laboratory of Environmental and Industrial Acoustics and Acoustic Comfort,
Federal University of Paraná, Brazil

Abstract

This chapter discusses the importance of evaluating the exposure to vibration, particularly whole-body vibration, of workers whose activities place them in automotive vehicles. To this end, we describe the procedures for measuring whole-body vibration established by the ISO standards ISO 2631-1:1997 and ISO 2631-5:2004, including the main descriptors, tolerance limits proposed by the standards, and the effects of vibration on human health and comfort. Lastly, a case study is presented to offer a practical example of how to measure whole-body vibration and analyze the results.

1. Introduction

The work environments of some professions involve an occupational hazard that is considered a possible aggravation factor in health, i.e., the levels of vibration to which the human body is exposed. According to Hill et al. (2009), vibration is considered one of the most common occupational hazards in industry worldwide and is directly linked to the prevalence of back pain in various professions. The literature states that other problems besides pain may also be ascribed to prolonged exposure to vibration, such as declining job performance and early onset of fatigue (Tiemessen et al., 2007).

[*] Email: bruno_sergio_por@yahoo.com.br.

Human sensitivity to vibration involves several ergonomic factors such as posture, muscle tension in the region of the back, and physical factors such as the frequency, direction and duration of the level of vibration (Mansfield et al., 2006). Thus, prolonged exposure to vibration can cause increased rates of absenteeism and become a problem for both workers and employers (Widanarko et al., 2012). It is therefore important to know the vibration levels to which employees are exposed in their work environment in order to make decisions about the alignment of these levels with technical standards.

2. International Standards for Whole-Body Vibration

Vibration in the human body is defined as any vibration generated by an external event and acting upon the body, as when the body is on an oscillating surface. Several occupations may generate vibrations in the human body, including the transportation of freight and people (Griffin, 1990).

According to Soeiro (2011), the study of vibrations can be divided into the following two main areas:

Whole-Body Vibrations

Vibrations that are transmitted to the entire body through a supporting surface, they be transmitted from the region of the feet, back and buttocks in a sitting position. These vibrations, which are of low frequency and high amplitude in the range of 1 to 20 Hz, are specific to transportation activities in vehicles such as trucks, tractors, forklifts, buses, trains, airplanes and boats. For purposes of standardization, they are measured according to the following international standards:

ISO 2631-1:1997: a standard that establishes the basic procedures for measuring and evaluating whole-body vibration.

ISO 2631-2:2003: a standard that provides guidelines to measure whole-body vibration in buildings.

ISO 2631-4:2001: a standard that provides guidelines to evaluate whole-body vibration transmitted to passengers in motor vehicles, with special focus on the sensation of comfort.

ISO 2631-5:2004: a standard that assesses the impact of time, vibration dose and multiple shocks on the health of workers.

Hand-Transmitted Vibration

This type of vibration reaches the upper limbs, mainly the hands and arms, and ranges from 6.3 to 1250 Hz. It is generated when working with hand tools (jackhammers, sanders, chainsaws, etc.) and is standardized by ISO 5349-1:2001 and ISO 5349-2:2001.

3. ISO 2631-1

According to Balbinot (2001), the criteria relating vibration exposure limits to health and safety factors in the population were initially proposed in 1967 and were subsequently incorporated into the ISO 2631 standards. The purpose of ISO 2631-1:1997 is to define methods that quantify the levels of whole-body vibration with respect to human health, comfort and perception of fatigue. The latest edition of 1997 features discussions about the evolution of experimental studies on the effects of whole-body vibration on health.

The ISO 2631-1:1997 standard can be used to assess whole-body vibrations generated by land, air and water vehicles (Okunribido et al., 2007; Oliveira et al., 2001; Asmussen et al., 2001). The appendixes of this standard provide information about the possible effects of vibration on health (Appendix B), comfort and perception (Appendix C) and the occurrence of nausea caused by vibratory movements (Appendix D).

ISO 2631-1:1997 states that vibration should be measured according to a system of coordinates that originates at the point where the vibration enters the human body. Three axes are evaluated in the seated, standing or recumbent positions: the x-axis (anteroposterior direction), y-axis (lateral direction) and z-axis (longitudinal direction). Figure 1 illustrates the location of the axes for the evaluation of whole-body vibration, showing the three positions the human body can adopt in space.

Vibrations transmitted to the human body should be measured at the interface between the body and the vibrating surface. The standard ISO 2631-1:1997 adopts three main areas for measuring vibration in people in the seated position: the surface of the seat, the backrest and the feet. Measurements at the seat surface should be taken under the coccyx (beneath the buttocks), while at the backrest they should be taken in the main area that supports the back, and at the feet they should be taken on the surface closest to where the feet are placed. For the prostrate position, the ISO 2631-1:1997 standard specifies measuring vibrations at the surface supporting the pelvis, back and head.

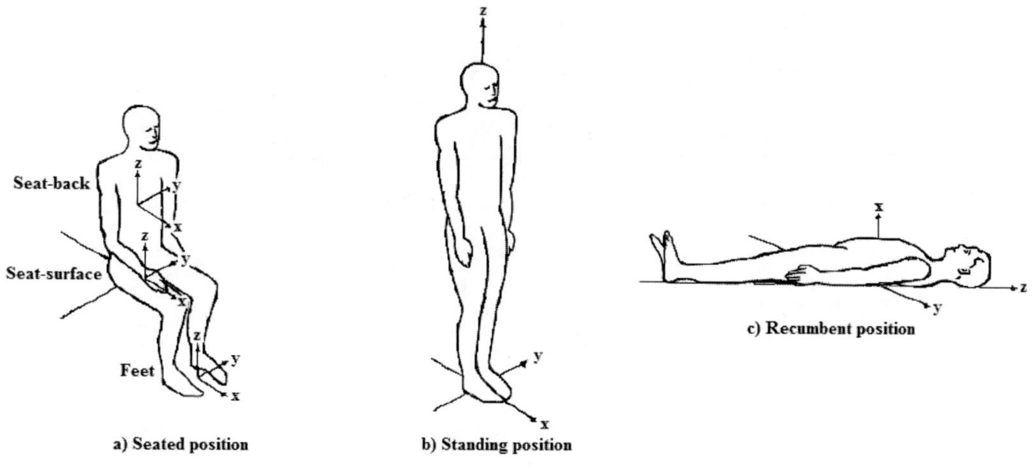

Source: ISO 2631-1:1997.

Figure 1. Basicentric axes of the human body.

The vibration measurement transducer normally used is a piezoelectric accelerometer, which captures the oscillations and transforms them into levels of acceleration. Two types of accelerometers are currently used: uniaxial accelerometers, which measure vibration on only one axis of the system of coordinates, and triaxial accelerometers, which measure vibration simultaneously on the three axes of the coordinate system.

To establish guidelines for the sensation of comfort in a vibrating environment, the ISO 2631-1:1997 standard presents rms acceleration values that indicate the reaction of people exposed to vibration effects and their subjective comfort response. The values listed in Table 1 show approximate indications of likely reactions of people exposed to vibrations.

With regard to possible health effects in workers exposed to whole-body vibration, the ISO 2631-1:1997 standard provides health guidance caution zones for 4 to 8 hours of exposure. Figure 2 shows health effects guidelines; the hachured area indicates is a potential health risk in this range of acceleration with respect to exposure time.

Table 1. rms acceleration related to comfort

rms acceleration ($m.s^{-2}$)	Classification
< 0.315	Not uncomfortable
0.315 to 0.63	A little uncomfortable
0.5 to 1.0	Fairly uncomfortable
0.8 to 1.6	Uncomfortable
1.25 to 2.5	Very uncomfortable
>2	Extremely uncomfortable

Source: ISO 2631-1:1997.

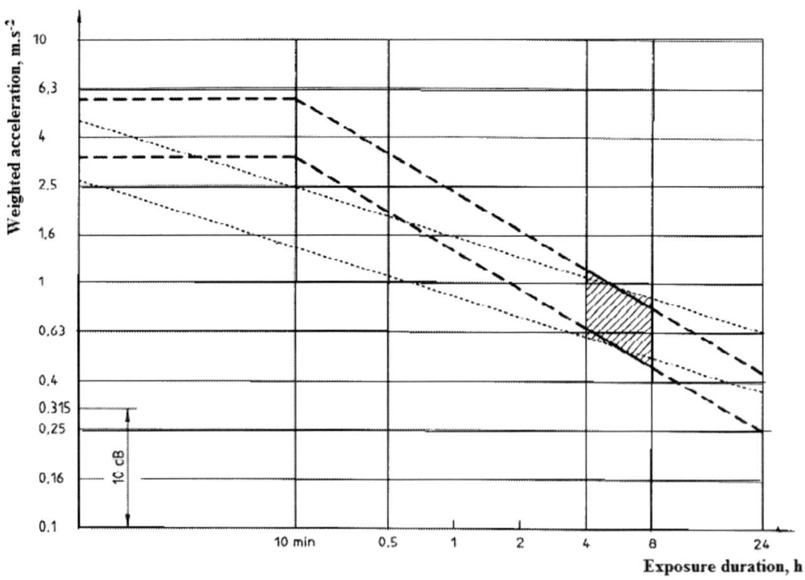

Source: ISO 2631-1:1997.

Figure 2. Health effects guidelines.

According to ISO 2631-1:1997, for a daily exposure of 8 hours, the graphic should be read as follows:

- The region of the curve up to 0.43 $m.s^{-2}$ indicates that health effects have not been documented clearly or observed objectively;
- The region of the curve from 0,43 $m.s^{-2}$ a 0.78 $m.s^{-2}$ indicates that precautions should be taken with respect to potential health risks; and
- The region of the curve above 0.78 $m.s^{-2}$ indicates probable health risks.

To complement the assessment of whole-body vibration, other reference values are given in Directive 2002/44/EC of the Council of the European Union. This document establishes the minimum requirements to protect workers against safety and health risks resulting from exposure to mechanical vibration. Thus, the directive specifies a value of daily exposure that should trigger some action, which is set at 0.5 $m.s^{-2}$ rms or 9.1 $m.s^{-1.75}$ VDV (Vibration Dose Value) and a daily exposure limit value of 1.15 $m.s^{-2}$ rms or 21.0 $m.s^{-1.75}$ VDV, both normalized for a reference period of eight hours of daily exposure.

4. Key Terms Used in the Measurement of Whole-Body Vibration

The measurement of the intensity of human exposure to vibration is related to the extent of oscillation and can be quantified by verifying the parameters of induced displacement, imposed velocity or transmitted acceleration, according to ISO 2631-1:1997. However, the latter physical magnitude appears to represent the intensity of the vibration more faithfully than the first two (Melo, 2006). Thus, the measure is analyzed based on acceleration data expressed in $m.s^{-2}$.

To evaluate whole-body vibration, the ISO standard 2631-1:1997 establishes the parameters that should be measured using accelerometers. A measurement may involve first checking the peak vibration, which represents the maximum acceleration in an interval of 1 second, measured in $m.s^{-2}$. Next, one of the key descriptors is the rms value, which represents the value of the average energy of the vibratory motion.

$$a_w = \left[\frac{1}{T}\int_0^T a_w^2(t)\,dt\right]^{\frac{1}{2}} \tag{1}$$

where:

a_w is the acceleration as a function of time, in [$m.s^{-2}$], and
T is the duration of the measurement, in [s].

Vibration Dose Value – VDV

The Vibration Dose Value, which is measured in 1 $m.s^{-1.75}$, is a cumulative measure that is used to reflect the total exposure of a worker exposed to vibration. The VDV serves to calculate the maximum vibration level for a given exposure, based on a period of time (usually 8 hours). The formula below shows the formulation of the VDV:

$$VDV = \left[\int_0^T a_W(t)^4 \, dt \right]^{\frac{1}{4}} \tag{2}$$

where:

$a_W(t)$ is the instantaneous acceleration, in [$m.s^{-2}$], and
T is the duration of the measurement, in [s].

Maximum Transient Vibration Value – MTV

The maximum vibration transmitted transiently is considered, which is defined as:

$$MTV = \max[a_w(t_0)] \tag{3}$$

where:

a_W is the highest magnitude of $a_W(t_0)$ read during the measurement, in [$m.s^{-2}$].

Crest Factor – CF

The crest factor (CF) is defined as the ratio between the peak acceleration value and the rms value measured in the same direction in a period of 1 minute. If the CF is above 9, additional methods of evaluation are recommended, since the measure is characterized by a high occurrence shocks.

$$CF = \frac{Peak}{rms} \tag{4}$$

where:

Peak is the maximum instantaneous value of peak acceleration, in [$m.s^{-2}$], and
rms is the value of acceleration during the period of time, in [$m.s^{-2}$].

The graph in Figure 3 depicts a measurement of whole-body vibration, showing the rms, MTV and peak ($m.s^{-2}$) values for evaluation of the z-axis within a given time interval.

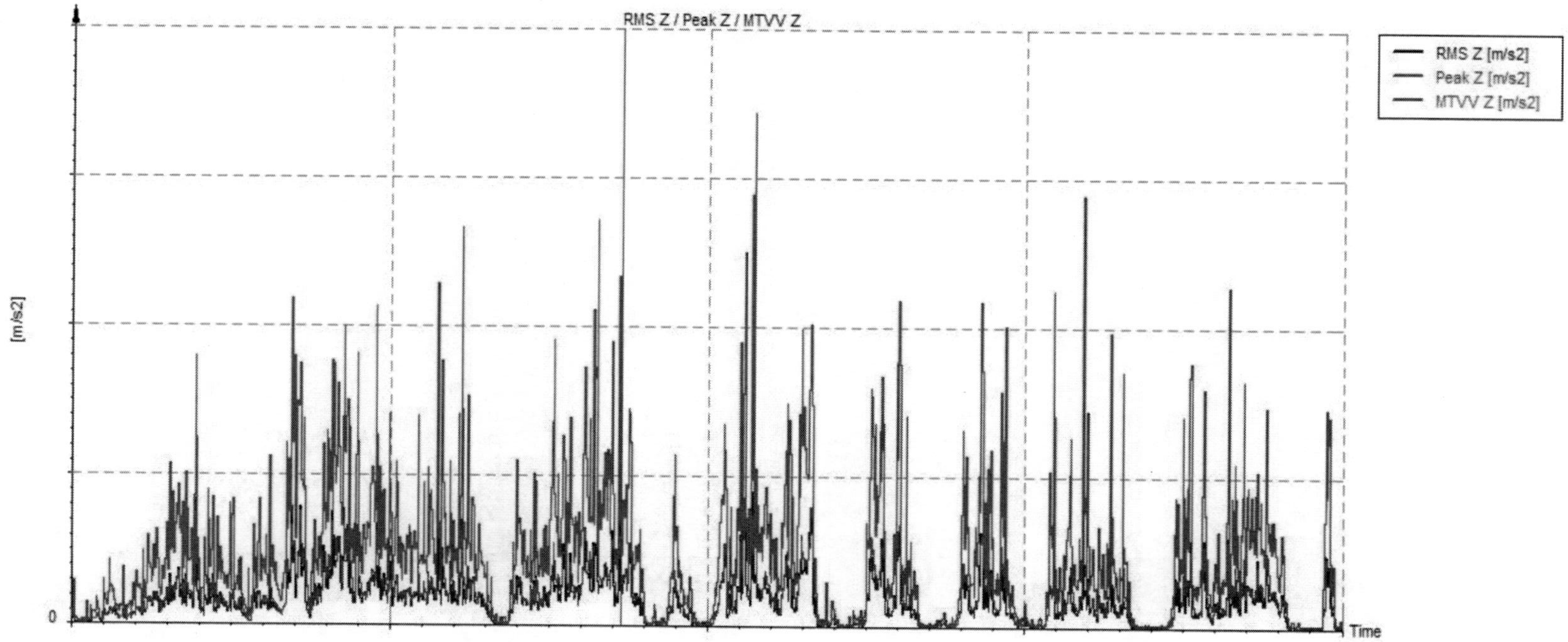

Figure 3. Measurement of whole-body vibration.

Weighted Vibration Total Value – VTV

The values of rms acceleration or dose obtained on the three axes can be added in order to obtain the total value or the value resulting from vibration. For purposes of standardization, ISO 2631-1:1997 uses this descriptor as the main indicator to evaluate the exposure of the human body to a given vibration level in an interval of time.

$$VTV = \left(k_x^2 . a_{wx} + k_y^2 . a_{wy} + k_{xz}^2 . a_{wz}\right)^{\frac{1}{2}}. \qquad (5)$$

where:

a_{wx}, a_{wy} and a_{wz} are the rms accelerations relative to the orthogonal axes x, y, z

and

k_x, k_y and k_z are the multiplication factors, x = 1.4, y = 1.4 and z = 1.

5. ISO 2631-5

The ISO standard 2631-5:2004 is indicated to assess human exposure to multiple mechanical impacts measured on the seat when the person is seated. According to this standard, the adverse health effects caused by prolonged exposure are related with vibration dose measurements obtained from acceleration peaks. These are determined using a method to the 6th power to verify that peaks with magnitudes below 1/3 of the maximum peak do not significantly influence the value of the dose.

Acceleration of Dose Calculation

The acceleration of dose, D_k, in m.s^{-2}, in the k direction is defined as:

$$D_k = \left[\sum_i A_{ik}^6\right]^{\frac{1}{6}} \qquad (6)$$

where:

A_{ik} is the ith peak acceleration response [m.s^{-2}] and
k = x, y or z.

The effects on health are determined based on the average daily dose, D_{kd}, in m.s^{-2}, to which a person can be exposed, using the following equation:

$$D_{kd} = D_k \cdot \left[\frac{t_d}{t_m}\right]^{\frac{1}{6}} \tag{7}$$

where:

t_d is the duration of daily exposure, in [s], and
t_m is the time interval in which D_k was measured, in [s].

6. Assessment of Health Effects

According to a biomechanical model proposed by ISO standard 2631-5:2004, which is based on experimental data, it has been shown that there is a linear relationship between the compressive stress on the lumbar spine and the peak acceleration of vibration. The static compressive stress, S_e, in megapascals is shown below:

$$S_e = [(m_k D_k)^6] \tag{8}$$

where:

D_k is the acceleration of dose in the k direction, in [m.s^{-2}].
The recommended values of m_k are:

m_x = 0.015 in [MPa/m.s^{-2}],
m_y = 0.035 in [MPa/m.s^{-2}], and
m_z = 0.032 in [MPa/m.s^{-2}].

The daily equivalent static compression dose, S_{ed}, is obtained using the formula:

$$S_{ed} = \left[\sum_{k=x,y,z} (m_k D_k)^6\right]^{\frac{1}{6}} \tag{9}$$

An equivalent static compression dose, S_{ed}, which exceeds 0.5 can be considered a risk factor for spinal problems. If the value exceeds 0.8, the ISO 2631-5:2004 standard warns that there is a potential risk for the appearance of neurodegenerative disorders, particularly in the region of the vertebral column.

The adverse health effects caused by the dose stimulus of vibration exposure as a result of accumulated exposure time are evaluated by calculating the R factor, as indicated in the following formula:

$$R = \left[\sum_{i=1}^{n} \left(\frac{S_{ed} . N^{\frac{1}{6}}}{S_{ui} - c} \right)^{6} \right]^{\frac{1}{6}} \tag{10}$$

where:

N is the number of days of exposure per year, in [days],
i is the year of the count,
n is the number of years of exposure, in (years),
c is the constant that represents the static stress due to gravitational force, in the seated position while driving vehicles, c = 0.25, in [MPa];
S_{ui} is the maximum force on the lumbar spine of a person aged (b+1) years; and
b is the age when exposure started, in (years).

The value of S_{ui}, in MPa, varies according to vertebral bone density, which usually decreases with age. The formula below shows the relationship between S_{ui} and b+1(in years):

$$S_{ui} = 6.75 - 0.066(b + i) \tag{11}$$

where:

i is the year when the count was taken, in [years], and
b is the age when exposure began, in [years].

ISO standard 236-5:2004 shows that values of R < 0.8 indicate a low probability of adverse health effects, while R > 1.2 indicates a high probability of negative health effects being felt. Thus, the ISO 236-5:2004 standard offers a method that complements ISO 2631-1:1997, correlating the vibration dose with the annual exposure time and establishing a standard value to quantify the possible effects of vibration on the human body.

7. CASE STUDY

To demonstrate the application of these standards to a real situation of vibration exposure, a case study is presented of a real measurement, the level of vibration exposure experienced by an urban bus driver. At the time of measurement in the year 2012, the driver was 30 years old, and had been driving buses for 5 years, on an average workday of eight hours.

The vibration level to which a driver is exposed is considered whole-body vibration and is measured as such. This measurement was taken using a type 4515-B-002 Triaxial DeltaTron Seat Pad Accelerometer and a Bruel and Kjaer 4447 Human Vibration Analyzer positioned on the driver's seat, in conformance with ISO 2631-1:1997 (see Figure 4). The driver was then asked to drive the bus on a normal workday route, to represent the period of time of a daily work cycle.

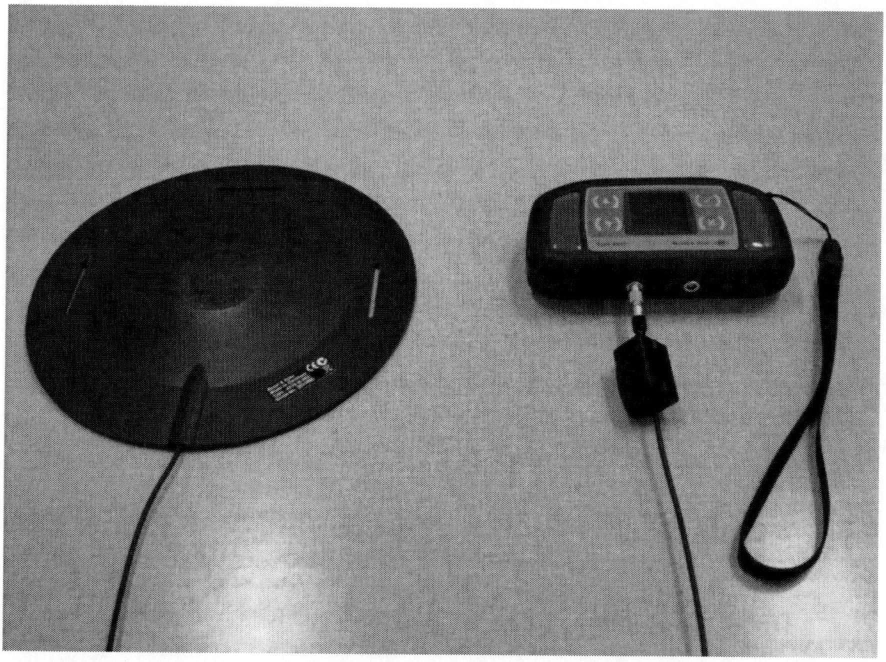

Figure 4. Triaxial DeltaTron Seat Pad Accelerometer (left) and Human Vibration Analyzer (right).

Following these instructions, the driver traveled along a normal bus route on an asphalt road for a total work cycle of 30 minutes. The whole-body vibration was measured in this 30-min period, representing a cycle that is repeated throughout the driver's workday, based on which the magnitude of the vibration level in a complete 8-hour workday can be estimated. After concluding the measurement, the collected data were downloaded into the Bruel and Kjaer BZ-5623 Vibration Explorer software program. Table 2 lists the main measured values .

Table 2. Values of whole-body vibration measured in an urban bus

Parameter	Unit	X	Y	Z	VTV
rms	$m.s^{-2}$	0.34	0.46	0.55	0.97
MTV	$m.s^{-2}$	1.08	1.35	1.76	
VDV	$m.s^{-1.75}$	4.43	4.99	7.16	12.65
Peak	$m.s^{-2}$	2.10	2.40	3.88	
CF		6.18	5.22	7.05	

The values of rms acceleration on the three axes, and the values of MTV and VDV, as well as the sum of VTV + rms and VTV + VDV were calculated automatically by the vibration analyzer software program. This make it easier to understand the values found in this stage of the evaluation, enabling faster decision-making by the evaluator.

The value of VTV is the weighted vector sum of the three axes ($0.97\ m.s^{-2}$) and, compared with the value of $0.78\ m.s^{-2}$ of ISO 2631-1:1997 established by its health guidance caution zone (HGCZ) limits, exceeds the recommended value. The rms acceleration of the major axis (z = 0.55 ms-2) characterized an "uncomfortable" work environment, according to Appendix C of the ISO 2631-1:1997 standard. Moreover, the value of the vector sum of the weighted vibration dose ($12.65\ m.s^{-1.75}$) exceeded the limit ($9.1\ m.s^{-1.75}$) specified by Directive 2002/44/EC of the Council of the European Union.

Based on the results of the measurement and a comparison of the values specified by the ISO 2631-1:1997 standard and by Directive 2002/44/EC of the Council of the European Union, it can be stated that the driver's work environment is conducive to serious health risks. Therefore, steps should be taken to reduce and control exposure to vibration in this vehicle in order to avoid future health problems.

The ISO 2631-5:2004 standard can be used to correlate vibratory environments with the occurrence of vertebral column disorders. This standard lists the values of vibration exposure that may lead to spinal injuries, which are calculated according to the steps described below.

First one computes the value of the acceleration of dose based on the peak acceleration values, using formula (6), which results in:

$$D_{x\ 30\ min} = 7.0\ m.s^{-2}$$
$$D_{y\ 30\ min} = 8.4\ m.s^{-2}$$
$$D_{z\ 30\ min} = 13.5\ m.s^{-2}$$

Considering an 8-hour workday, the daily dose is estimated using formula (7):

$$D_x = 7.0\left[\left(\frac{8*60}{30}\right)^{\frac{1}{6}}\right] = 11.11\ m.s^{-2}$$

$$D_y = 8.4\left[\left(\frac{8*60}{30}\right)^{\frac{1}{6}}\right] = 13.33\ m.s^{-2}$$

$$D_z = 13.5\left[\left(\frac{8*60}{30}\right)^{\frac{1}{6}}\right] = 21.43\ m.s^{-2}$$

Using these values, the equivalent static compression dose on the vertebral column of the worker (S_{ed}) can be estimated, considering: $m_x = 0.015\ MPa/m.s^{-2}$; $m_y = 0.035\ MPa/m.s^{-2}$; $m_z = 0.032\ MPa/m.s^{-2}$.

$$S_{ed} = [(0.015 * 11.11)^6 + (0.035 * 13.33)^6 + (0.032 * 21.43)^6]^{\frac{1}{6}}$$
$$S_{ed} = 0.70\ MPa$$

The result indicates a moderate effect on health ($0.5MPa < S_{ed} < 0.8MPa$) for this driver who is exposed to this vibration level during his workday.

Based on the age when exposure begins (b = 25 years) and the exposure time considered (i = 5 years), one can estimate the yield stress of the driver's lumbar spine (S_{ui}) for each year of exposure, as follows:

$$S_{ui\,(i+1)} = 6.75 - 0.066(25 + 1) = 5.03\ \text{MPa}$$
$$S_{ui\,(i+2)} = 6.75 - 0.066(25 + 2) = 4.97\ \text{MPa}$$
$$S_{ui\,(i+3)} = 6.75 - 0.066(25 + 3) = 4.90\ \text{MPa}$$
$$S_{ui\,(i+4)} = 6.75 - 0.066(25 + 4) = 4.84\ \text{MPa}$$
$$S_{ui\,(i+5)} = 6.75 - 0.066(25 + 5) = 4.77\ \text{MPa}$$

Using these data, the R factor can be calculated, as follows:

$$R = \left\{\left[\frac{0.7\left(240^{\frac{1}{6}}\right)}{5.03 - 0.25}\right]^6 + \left[\frac{0.7\left(240^{\frac{1}{6}}\right)}{4.97 - 0.25}\right]^6 + \left[\frac{0.7\left(240^{\frac{1}{6}}\right)}{4.90 - 0.25}\right]^6 + \left[\frac{0.7\left(240^{\frac{1}{6}}\right)}{4.84 - 0.25}\right]^6 + \left[\frac{0.7\left(240^{\frac{1}{6}}\right)}{4.77 - 0.25}\right]^6\right\}^{\frac{1}{6}}$$

R = 0.49

This result indicates a low probability of negative health effects on this worker in three years of exposure (R<0.8). However, since the evaluation based on the parameters of the ISO 2631-1:1997 standard indicated health risks, it is advisable to adopt measures aimed at reducing the vibration levels in the driver's workstation.

CONCLUSION

This chapter addressed the issue of exposure to whole-body vibration in occupational settings according to ISO 2631-1:1997, which describes measurement methods and to ISO 2631-5:2004, which estimates the magnitude of the effect of vibration on the human body, focusing mainly on the lumbar spine. Hopefully, the procedures described herein will be applied in practice to measure exposure to occupational vibrations, in order to underpin decisions regarding the adjustment of these vibrations to levels tolerated by the human body, thereby preserving the health of workers.

ACKNOWLEDGMENT

The authors gratefully acknowledge the German Government, through the German Academic Exchange Service – DAAD (Deutscher Akademischer Austauschdienst) and the Brazilian Government, through the National Council for Scientific and Technological

Development – CNPq, for their financial support, which enabled the purchase of the sound level meters and software used in this study.

REFERENCES

Asmussen I, Menzel W, Mumm H. *Ship Vibration*. Hambug: Gl-Group. 2001.

Balbinot A. *Caracterização dos níveis de vibração em motoristas de ônibus: um enfoque no conforto e na saúde* [Characterization of vibration levels in bus drivers: an approach to comfort and health]. Porto Alegre, 2001. Thesis (PhD).

Griffin MJ. *Handbook of human vibration*. London: Academic Press, 1990.

Hill TE, Desmoulin GT, Hunter C. Is vibration truly an injurious stimulus in the human spine? *Journal of Biomechanics*, 2009, 42 (16): 2631–2635.

International Organization for Standardization. ISO 2631-1: Mechanical vibration and shock - evaluation of human exposure to whole-body vibration. *Part 1: General requirements*. Switzerland, 1997.

International Organization for Standardization. ISO 2631-2: Mechanical vibration and shock - evaluation of human exposure to whole-body vibration. *Part 2: Vibrations in buildings (1Hz to 80Hz)*. Switzerland, 2003.

International Organization for Standardization. ISO 2631-4: Mechanical vibration and shock - evaluation of human exposure to whole-body vibration. *Part 4: Guidelines for the evaluation of the effects of vibration and rotational motion on passenger and crew comfort in fixed-guide way transport systems*. Switzerland, 2001.

International Organization for Standardization. ISO 2631-5: Mechanical vibration and shock - evaluation of human exposure to whole-body vibration. *Part 5: Method for evaluation of vibration containing multiple shocks*. Switzerland, 2004.

Mansfield NJ. *Human response to vibration*. Boca Raton: CRC Press, 2004.

Melo RBM. *Exposição ocupacional a vibrações transmitidas ao corpo inteiro: factores condicionantes na condução de autocarros urbanos* [Occupational exposure to whole-body vibrations: conditioning factors in driving urban buses]. Lisbon, 2006. Thesis (PhD).

Okunribido OO. Shimbles SJ, Magnusson M, Pope M. City bus driving and low back pain: A study of the exposures to posture demands, manual materials handling and whole-body vibration. *Applied Ergonomics,* 2007, 38 (1): 29-38.

Oliveira CG, Simpson DM, Nadal J. Lumbar back muscle activity of helicopter pilots and whole-body vibration. *Journal of Biomechanics,* 2001, 34 (10): 1309-1315.

Soeiro NS. *Vibrações e o corpo humano: uma avaliação ocupacional* [Vibrations and the human body: an occupational evaluation]. *Acústica e vibrações* [Acoustics and vibrations], 2001, 43: 1-10.

Tiemessen IJ, Hulshof CTJ, Frings-Dresen MHW. An overview of strategies to reduce whole-body vibration exposure on drivers: A systematic review. *International Journal of Industrial Ergonomics*, 2007, 37 (3): 245-256.

Widanarko B, Legg S, Stevenson M, Devereux J, Eng A, Mannetje A, Cheng S, Pearse N. Prevalence and work-related risk factors for reduced activities and absenteeism due to low back symptoms, *Applied Ergonomics*, 2012, 43: 727-737.

Wasserman DE, Wilder DG, Pope MH, Magnusson M, Aleksiev AR, Wasserman JF. Whole-body vibration exposure and occupational work-hardening. *Journal of Occupational and Environmental Medicine*, 1997, 39 (5):403-407.

In: Noise and Ergonomics in the Workplace
Editor: Paulo Henrique Trombetta Zannin

ISBN: 978-1-62417-573-2

Chapter 12

CHARACTERIZATION AND ANALYSIS OF OCCUPATIONAL NOISE: CASE STUDY OF A SERVICE COMPANY OF THE PUBLIC TRANSPORT SECTOR

Elaine Carvalho da Paz [*] ***and Paulo Henrique Trombetta Zannin***
Laboratory of Environmental and Industrial Acoustics and Acoustic Comfort (LAAICA),
Federal University of Paraná, PR, Brazil

ABSTRACT

Noise is one of the most nocuous agents in the workplace. Evaluating its dissemination in social and work environments becomes increasingly important when one considers that the injuries it causes are irreversible, e.g., auditory injuries, and that exposure to noise pollution leads to other physical and psychological disorders that negatively affect general health and quality of life. Despite these deleterious effects, the problem of occupational noise in small businesses is still little explored. This chapter describes a case study of a service company of the public transport sector, analyzing occupational noise exposure in the workplace and underpinning the analysis with national and international regulations for occupational health procedures. Each workstation was assessed objectively based on sound pressure level and noise dose measurements, and subjectively based on questionnaires answered by employees. A canonical correlation analysis confirmed the existence and intensity of correlation between the two datasets obtained from the subjective and objective assessments.

1. INTRODUCTION

The advent of new technologies and the organization of industrial work processes have led to changes in the social model. From the standpoint of occupational health, it is clear that exposure to excessive noise levels associated with work shifts, high levels of responsibility,

[*] pazelaine@hotmail.com; epaz@bol.com.br.

forced attention, fatigue and health conditions mean that many industrial environments are still far from offering ideal working conditions (Rothman and Greenland, 1998; Morata and Dunn, 1995; Dwyer, 1991).

Since people spend about 75% of their time in work environments, the management of these spaces should necessarily focus on preserving the physical and psychological integrity of the individuals working in them (Lamb et al., 2005; Adera et al., 1993; Kjellberg, 1990).

Despite the deleterious effects of noise pollution on human health and the increase in work-related health problems (Alberti, 1998), the issue of occupational noise in small businesses is still neglected in developing countries such as Brazil, through the establishment of obsolete industrial plants and noncompliance with occupational hygiene and health (Barreto et al., 1997).

In the analysis of occupational noise, the chosen research strategy should allow for correlations to be made of the organic effects that are felt and the perception of each employee, i.e., subjective evaluations. These subjective effects should be confirmed based on values measured in the workstation, i.e., through objective assessments, which provide evidence of the existence of potential negative effects on workers' health (Berglund et al., 1999; Berglund and Lindvall, 1995; Melamed et al., 1994; Melamed et al., 1992).

2. Standards for the Evaluation of Occupational Noise

The Regulatory Standards [Normas Regulamentadoras – NR] of Brazil's Ministry of Labor and Employment establish guidelines for managing occupational health in the country (MTE, 1978). Standards NR-07, NR-09, NR-15 and NR-17 set down specific guidelines for industrial noise (MTE, 1978).

With regard to occupational noise exposure, the NR-07 standard establishes the mandatory application of air conduction pure tone audiometric tests, independently of the use of personal protective equipment (MTE, 1978).

The NR-09 standard establishes the mandatory preparation and implementation of Environmental Risk Prevention Programs – ERPPs aimed at preserving workers' health and physical integrity, based on predictive, preventive and corrective evaluations of environmental risks, which encompass physical risks that include workers' exposure to occupational noise (MTE, 1978).

The ergonomic standard NR-17 (MTE, 1978) states that noise levels in work environments where constant intellectual demands and attention are required must comply with the Brazilian NBR 10152 standard for acoustic comfort (ABNT, 1987). If there is no equivalence or correlation with the activities listed in the NBR-10152 standard, the acceptable level of intermittent or continuous noise to ensure comfort should not exceed 65 dB(A).

Appendix 1 of the NR-15 standard (MTE, 1978) establishes the tolerance limits for sound pressure levels as a function of exposure time, as indicated in Table 1.

Occupational noise exposure can be assessed based on the equivalent continuous sound pressure level, L_{eq}, and the daily noise dose, D (WHO, 2012; Darlington, 2003; Berger et al., 2000).

Table 1. Admissible equivalent continuous sound pressure level, L_{eq}, as a function of daily noise exposure – according to Appendix 1 of the NR-15 standard

L_{eq} dB(A)	Maximum admissible daily noise exposure
85	8 hours
86	7 hours
87	6 hours
88	5 hours
89	4 hours, 30 minutes
90	4 hours
91	3 hours, 30 minutes
92	3 hours
93	2 hours, 40 minutes
94	2 hours, 15 minutes
95	2 hours
96	1 hour, 45 minutes
98	1 hour, 15 minutes
100	1 hour
102	45 minutes
104	35 minutes
105	30 minutes
106	25 minutes
108	20 minutes
110	15 minutes
112	10 minutes
114	8 minutes
115	7 minutes

The NR-15 standard defines the noise dose as a function of the equivalent noise level and exposure time, taking as reference a constant and continuous exposure to 85 dB(A) for 8 hours, which corresponds to a dose of 100%, with a 5 dB(A) doubling rate (MTE, 1978).

Figures 1 and 2 show the sound pressure level meter and noise dose meter for measuring the equivalent continuous sound pressure level, L_{eq}, and the noise dose, D, respectively.

If there is any period during the workday when noise levels exceed 115 dB(A), the situation will be characterized as a serious and imminent risk to employees that do not use personal protective equipment, whatever the final noise exposure dose may be (Fundacentro, 1999; OSHA, 1981). In this case, even if the dose is lower than 100%, which indicates that the tolerance limit has not been exceeded, the work activity must be stopped immediately. This type of assessment is performed pointwise during the workday of the worker (OSHA, 1981).

Table 2 describes the interpretation of noise dose values according to the Brazilian Occupational Hygiene Standard NHO-01 (Fundacentro, 1999).

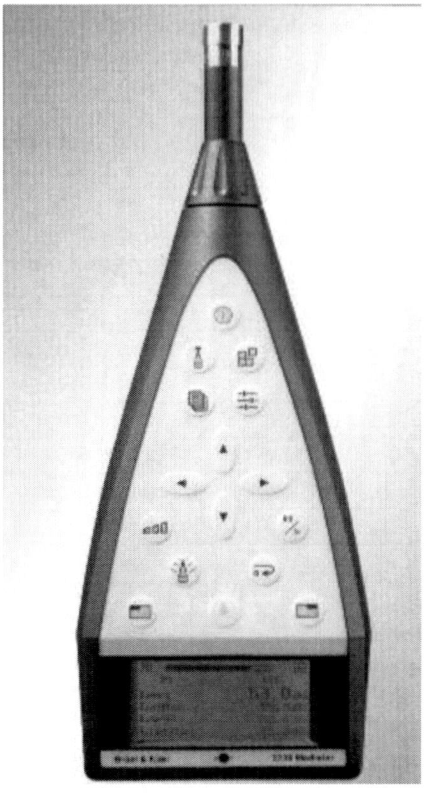

Figure 1. Sound pressure level meter – BK 2238 Mediator.

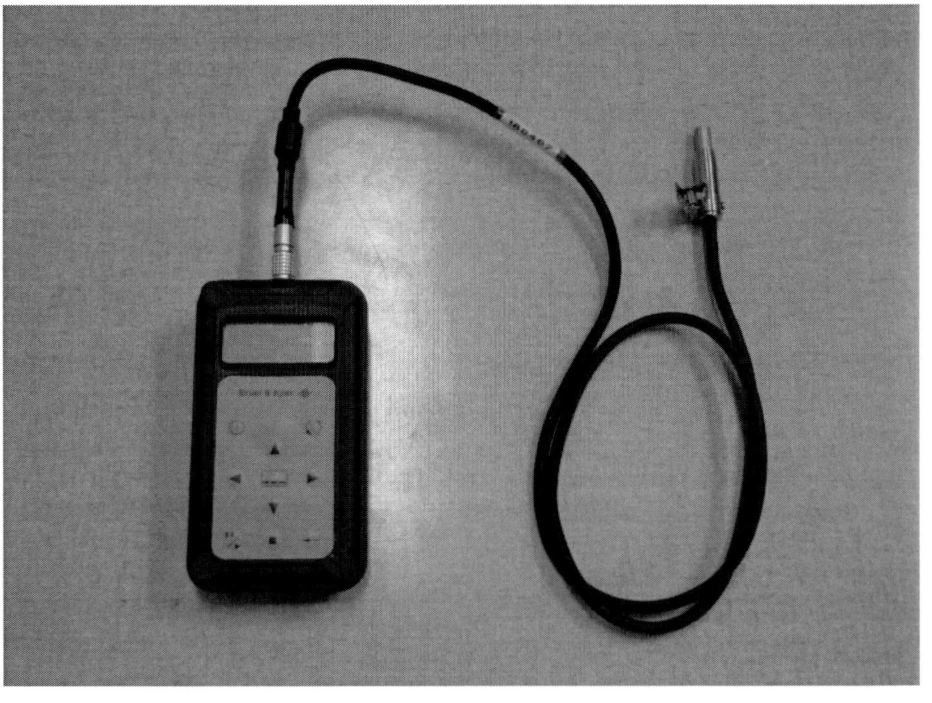

Figure 2. Noise dose meter – Brüel & Kjaer BK 4445.

Table 2. Brazilian Occupational Hygiene Standard NHO-01 – Interpretation of noise doses, D (Fundacentro, 1999)

Value of the dose D (%)	Maximum acceptable noise level dB(A)	Situation of exposure	Technical assessment of the situation	Activate control actions
10 to 50	80	Acceptable	------------	Desirable – not a priority
51 to 80	83	Acceptable	Deserves attention	Routine
81 to 100	85	Temporarily acceptable	Deserves attention	Preferential
101 to 300	92	Unacceptable	Deserves attention	Urgent
Above 301	115	Unacceptable	Emergency	Immediate
Any level	115	Unacceptable	Emergency	Immediate – Stop exposure

3. Case Study of a Service Company of the Public Transport Sector

This chapter discusses occupational noise assessment at a service company of the public transport sector. The analysis of occupational noise considered different work functions and different work shifts – day and night. For a more effective diagnosis of the work environment, objective and subjective assessments were made, the former by measuring sound pressure levels and noise dose, and the latter by ascertaining the workers' perception of noise in their work environment.

The objective evaluations involved taking two measurements at each workstation during the day and night shifts. These measurements were taken with a BK 2238 Mediator sound pressure level meter (see Figure 1), which meets the standards of the International Electrotechnical Commission – IEC and the American National Standards Institute – ANSI (IEC, 1979, 1985, 1988; ANSI, 1983). The noise dose was measured with a BK 4445 noise dose meter (see Figure 2), which meets the standards of ANSI (ANSI 1991). The measurements were taken based on the following premises: 1) the workstations were previously evaluated to define measuring points in each of them; 2) the measuring time at each point was limited to three minutes (Portela and Zannin, 2010).

Among the workstations selected for the measurements, in addition to the main workstation, secondary workstations that affect the worker during his workday, through direct or indirect contact, were also analyzed.

Tables 3 and 4 describe the workstations analyzed, as well as the values of equivalent continuous sound pressure level and the daily noise dose measured during the day and night periods.

Table 3. Occupational noise assessment – Nighttime

Function	Main workstation	L_{eq} dB(A)	Exposure time (%)	Dose D (%)	Secondary workstation	L_{eq} dB(A)	Exposure time (%)	Dose D (%)
Mechanic	Assembly line	83.7	62.5	46	Rubber workshop	81.1	19.4	7.98
					Lubrication	75.8	6.25	0.74
					Welding	70.1	6.25	0.2
Washer	Lubrication	75.8	69.4	8.26	Washing	83.3	12.5	0.085
					Welding	70.1	12.5	0.004
Lubricator	Lubrication	75.8	62.5	7.44	Washing	83.3	6.25	4.19
					Welding	70.1	6.25	0.2
					Assembly line	83.7	19.4	14.3
Welder	Welding	70.01	81.9	2.66	Assembly line	83.7	12.5	9.21
Tire fitter	Rubber workshop	81.1	75	30.9	Assembly line	83.7	19.4	14.3
Controller	Assembly line	83.7	94.4	69.5				
Electrician	Assembly line	83.7	81.9	60.3	Rubber workshop	81.1	12.5	5.15

Table 4. Occupational noise assessment – Daytime

Function	Main workstation	L_{eq} dB(A)	Exposure time (%)	Dose D (%)	Secondary workstation	L_{eq} dB(A)	Exposure time (%)	Dose D (%)
Mechanic	Assembly line	81.9	62.5	30.6	Rubber workshop	75.3	19.4	2.06
					Lubrication	74.4	6.25	0.54
					Welding	82.8	6.25	3.8
Washer	Lubrication	81.5	69.4	30.8	Washing	74.4	12.5	1.09
					Welding	82.8	12.5	7.6
Lubricator	Lubrication	74.4	62.5	5.43	Washing	81.5	6.25	2.78
					Welding	82.8	6.25	3.8
					Assembly line	81.9	19.4	9.49
Welder	Welding	82.8	81.9	49.8	Assembly line	81.9	12.5	6.12
Tire fitter	Rubber workshop	75.3	75	7.98	Assembly line	81.9	19.4	9.49
Controller	Assembly line	81.9	94.4	46.2				
Electrician	Assembly line	81.9	81.9	40.1	Rubber workshop	75.3	12.5	1.33
Lathe operator	Lathe	78.4	75	16.5	Assembly line	81.9	12.5	6.12
					Welding	82.8	6.88	4.18

Each individual has a unique perception of how noise pollution affects his daily life (Zannin, 2006). This subjective sensitivity may conceal the harmful effects of noise during the workday (Paz et al., 2005). Thus, for a more complete analysis of occupational conditions in the workplace, a subjective analysis was also made, using questionnaires, to ascertain the workers' perception of noise.

The questionnaires were filled out by the workers themselves during their workday, without the intervention of the researchers. Depending on the size of a company, subjective analysis should be performed by dividing the interviewees into groups according to age, gender, etc. In this case study, for example, the interviews involved 28 males from 27 to 51 years old, employed by a public transport company and working alternating shifts. A representative sample of employees was selected randomly among the various professional categories working day and night shifts.

Each questionnaire was answered anonymously and consisted of three parts. The first part contained questions about the worker's personal data (function, age, duration of employment, work shift, etc.). The second part of the questionnaire addressed the issue of occupational noise, and involved questions such as: The possibility of noise causing health problems; the nuisance caused by noise pollution; the environmental factor causing the greatest disturbance; the influence of noise generated in the work environment; classification of the noise generated in the workplace; and classification of the exposure to intense noise during the workday. The third part of the questionnaire, which addressed the use of hearing protection devices, included questions about the importance of using hearing protection; the use of the protector; the type of protector used; the discomfort resulting from its use; and the need for its shape to be modified to minimize the discomfort it causes.

3.1. Correlation between Objective and Subjective Assessments

There are several ways to analyze two datasets. One of the most common models of analysis is multiple regression, which involves explaining a variable by means of a linear combination of other variables (Hair et al., 1995).

Canonical correlation analysis can be considered an extension of multiple regression. The basic principle of canonical correlation is to develop a linear combination on each set of variables, thereby maximizing the correlation between the two sets (Johnson and Wichern, 1988). Canonical correlation analysis makes no distinction between independent and dependent variables, and there are only two sets of variables for which one seeks to identify the maximum correlation (Kendall, 1980).

Subjective and objective analyses were made to assess occupational noise, which generated two distinct datasets. In this case, canonical correlation was the most appropriate tool to correlate the two analyses.

The basic equation is expressed thus by Johnson and Wichern (1988):

$$Y_1 + Y_2 + Y_3 + \ldots + Y_n = X_1 + X_2 + X_3 + \ldots + X_n \qquad (2)$$

Where:

Y_n represent the variables of the set – objective assessment
X_n represent the variables of the set – subjective assessment

4. Results and Discussion

The objective analysis involved the interpretation of the sound pressure levels and noise dose measured in the main and secondary workstations:

1. During the night shift, the measurements in the assembly operation reached an equivalent sound level of 83.7 dB(A), which was slightly higher than the level of 81.9 dB(A) recorded in the daytime. Among all the employees, only the car washers were not affected by the noise of the assembly operation. The highest equivalent sound levels higher were measured during the night shifts. Only welders were subjected to a lower noise level at night, although they were also affected by the assembly operation. These employees suffered an increase of 6.8 dB(A) in noise pollution in their work environment, which was due to the fact that the welding operation showed an increase of 12.7 dB(A) over the noise level of 70.1 dB(A) measured at night.
2. The noise level of 73.3 dB(A) measured in the lathing operation during night shifts was attributed to the proximity of this workstation to the assembly operation, since the lathe operator only works the day shift. In this period, the noise level increased by 5.1 dB(A).
3. The less noisy environments such as washing, lubrication and rubber workshop are strongly affected by their noisier neighbors, such as assembly, lathing and welding. This causes these workstations to have high equivalent sound levels during both day and night shifts.

Table 5 describes the equivalent sound levels measured in the workstations, while Table 6 lists the equivalent sound levels to which these workers are exposed.

Table 5. Equivalent sound levels measured in the workstations

Workstation	L_{eq} dB(A)	
	Night	Day
Lathe	73.3	78.4
Assembly	83.7	81.9
Rubber workshop	81.1	75.3
Lubrication	75.8	74.4
Washing	83.3	81.5
Welding	70.1	82.8

Table 6. Equivalent sound levels in each function

Function	L_{eq} dB(A)	
	Night	Day
Tool controller	83.4	81.6
Electrician	83.2	81.2
Auto mechanic	82.4	80.7
Tire Fitter	81.5	77.4
Lubricator	79.2	78.3
Car washer	77.3	81.0
Welder	75.7	82.5
Lathe operator	0.0	79.3

Table 7. Daytime and nighttime noise dose

Function	Noise dose D (%)	
	Night	Day
Tool controller	69.5	46.2
Electrician	65.4	41.4
Auto mechanic	55	37
Tire fitter	45.2	17.5
Lubricator	26.1	21.5
Car washer	17	39.5
Welder	11.9	55.9
Lathe operator	0.0	26.8

In the daytime, only the dose measured for welders exceeded the regulated values, making the use of personal protective equipment indispensable. The other workers on day and night shifts were exposed to acceptable noise levels according to the NR 15 standard (see Table 1).

After analyzing the questionnaires for the subjective assessment, it was concluded that 80% of the company's employees are aware that noise pollution is harmful to health. However, only a small proportion claimed they felt some degree of discomfort due to the noise generated in their workstations. Even with noise levels of around 82 dB(A) measured in the workstations during the day shift and 83 dB(A) during the night shift, only 18% of the day shift employees and 7% of the night shift employees stated they experienced degree of discomfort due to noise pollution in the workplace (see Figure 3).

Tool controllers, electricians and car mechanics were found to be the workers most affected by noise during the night shift, while welders were the most affected by noise in the daytime. Because they work in the noisiest environments, electricians are exposed to higher sound pressure levels.

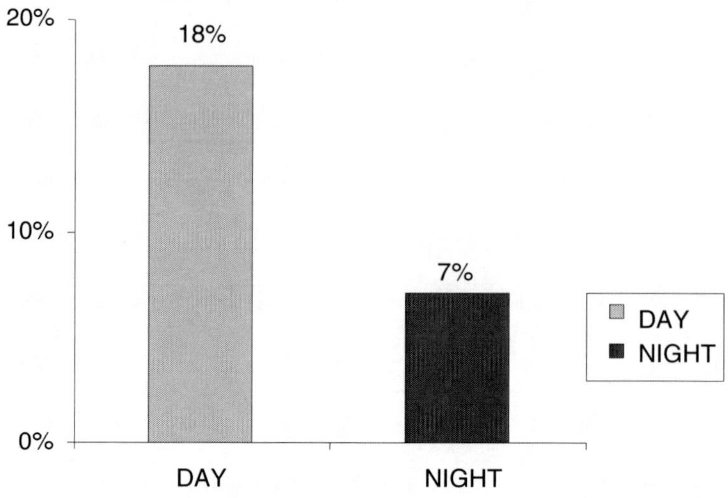

Figure 3. Discomfort caused by noise in the workplace as a function of daytime and nighttime shifts.

Table 7 describes the noise dose according to the worker's function. During night shifts, the workers most affected by noise are exposed to doses exceeding 50%. This exposure is characterized as deserving attention and requiring actions to control the situation (see Table 2).

The highest noise levels were measured at night. However, according to the subjective assessment, 45% of the employees working the night shift classified their perception of noise as "low intensity," and only 19% of night shift workers classified it as "very high." The day shift workers were more sensitive to noise, with 38% classifying their perception of noise generated in their workstations as "severe," and 31% as "very high" (see Figure 4).

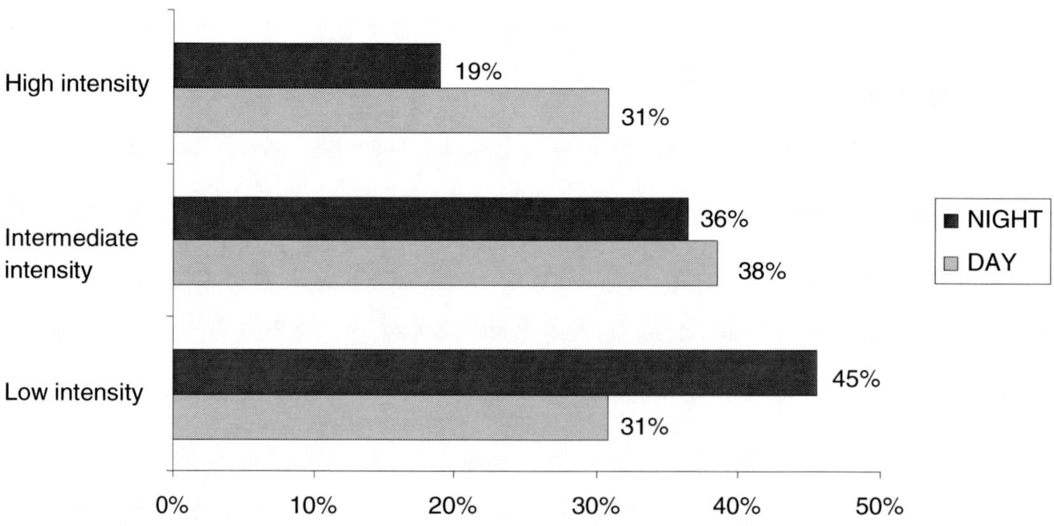

Figure 4. "How do you classify the relative intensity noise in your workstation?"

When asked about their perception of noise in neighboring workstations when their own workstation was not in operation, the majority of respondents from both shifts, i.e., 58% at night and 43% in the daytime, classified their noise exposure as "low intensity." This confirms the low perception of individuals about the influence of neighboring workstations on the noise pollution in their own workstation (see Figure 5).

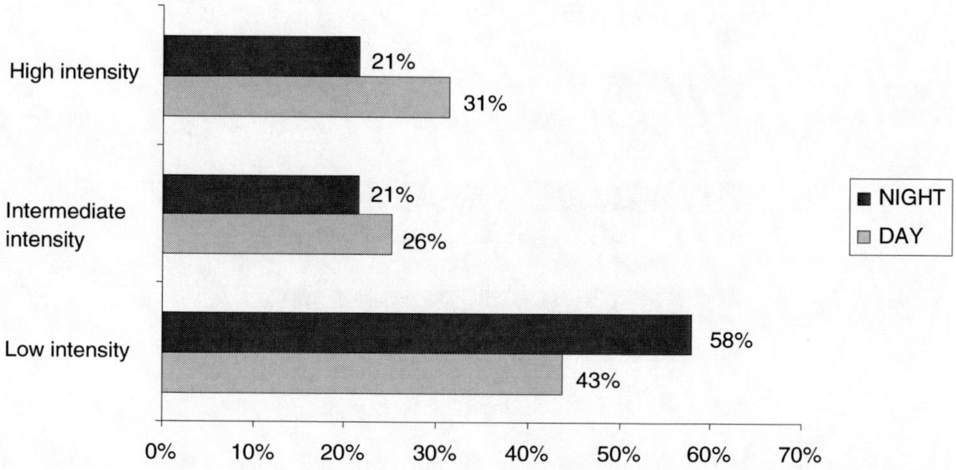

Figure 5. "How do you classify the relative intensity noise in neighboring workstations?"

Regarding the physiological effects experienced after continuous exposure to noise, 45% of respondents from the night shift and 43% from the day shift reported that the strongest effect was tinnitus (see Figure 6). With regard to the second strongest effect, differences were found between the two shifts. The night shift workers reported that the second strongest effect of noise was irritability (35% of respondents), while the day shift workers reported headaches (about 39% of respondents). It should be noted that, for this question, it was stipulated that a minimum of three and a maximum of seven effects could be reported by respondents, resulting in cross tabulations.

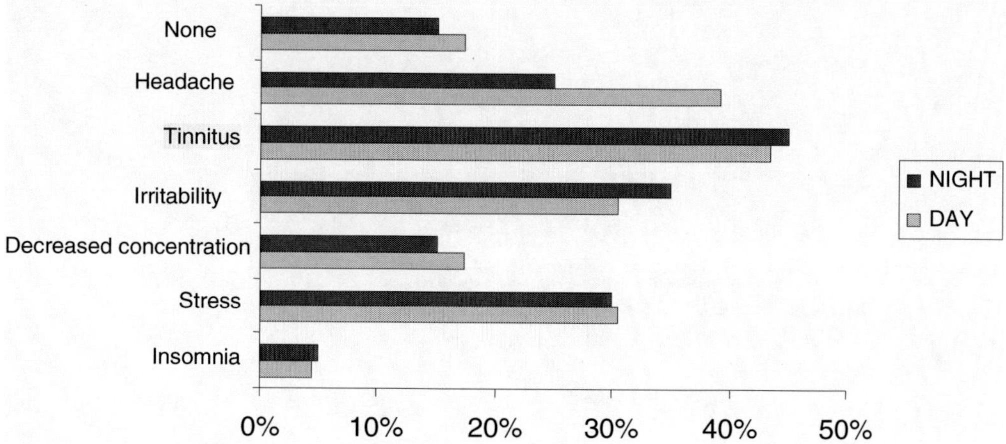

Figure 6. "What effects does noise have on you?"

When asked which factor in their work environment caused the greatest annoyance, 39% of the day shift workers and 35% of the night shift workers stated that it was noise pollution (see Figure 7).

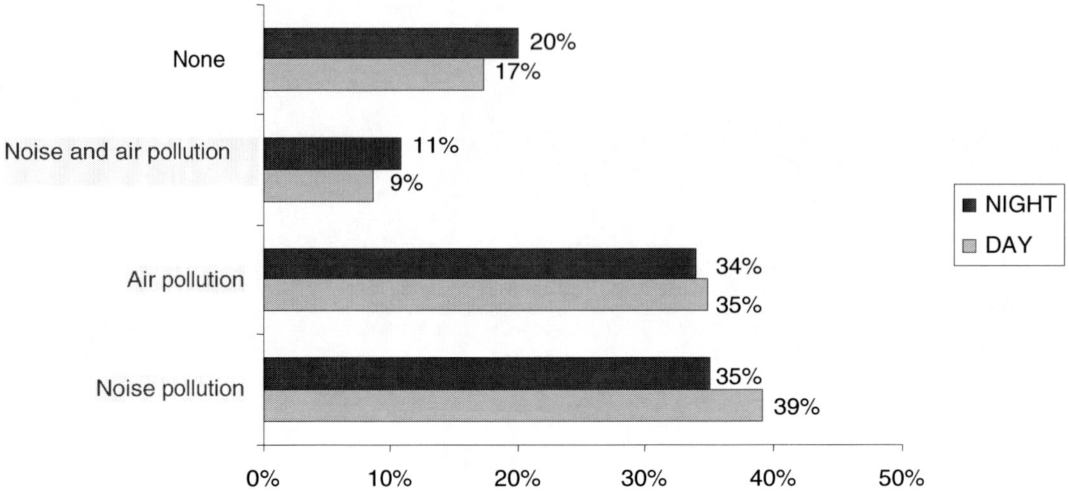

Figure 7. "What is the greatest factor of annoyance in your work environment?"

With regard to the use of hearing protectors, all the interviewees stated that they consider their use important, but only 75% of the respondents from the two shifts stated that they use hearing protectors on a regular basis.

Ear muffs are used alternately by 45% of the night shift workers and by 43% of the day shift workers. On the other hand, ear plugs are used alternately by 31% of the day shift workers and by 35% of the night shift workers. Only a small number of the respondents stated that they use the two hearing protectors simultaneously (see Figure 8).

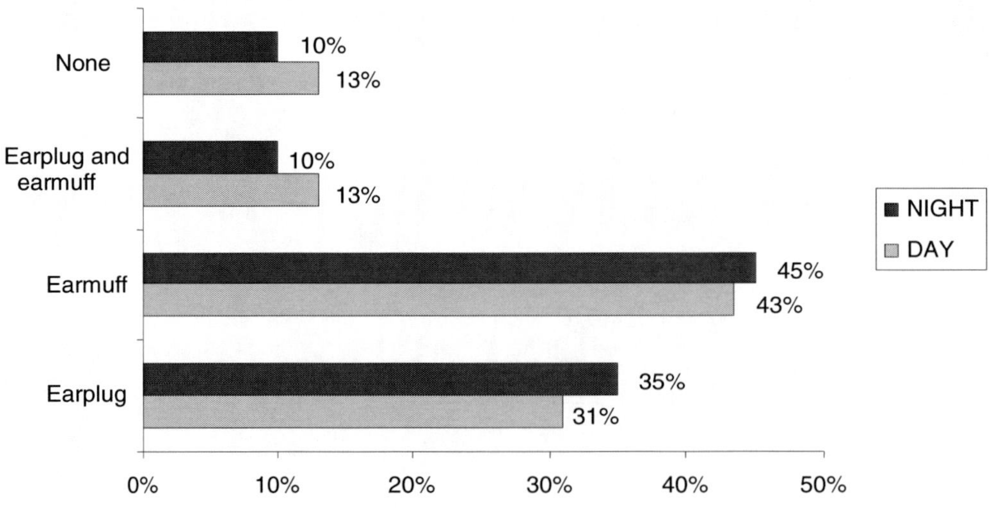

Figure 8. Type of hearing protector used during the workday.

About 43% of the day shift workers and 30% of the night shift workers find hearing protectors uncomfortable and would like their shape to be modified and improved.

The objective and subjective analyses were interpreted jointly by performing a canonical correlation analysis, which involved the following six steps:

1. Specification of the objectives of the analysis: The two datasets were tabulated as a group of objective data (L_{eq} and dose) and a group of subjective data (responses of the questionnaire applied). The objective of the modeling was when established: "What is the employees' level of awareness about the measured acoustic descriptors L_{eq} and the daily noise dose, D?"
2. Development of the plane analysis: The size of the sample to be used was determined. The minimum recommended size of the sample is 10 times the number of variables to be analyzed (Kendall, 1980).
3. Testing the assumptions of correlation: Each variable was tested based on the following assumptions: Is there isolated linear correlation and normality? Is there homoscedasticity and multicolinearity in the distribution of the data? (Johnson and Wichern, 1988). The results showed that there was a relationship between the two groups and that only the questions of the second and half of the third part of the questionnaire could be applied to the modeling.
4. Model estimation and calculation of the explanatory power: This step involved calculating the model from the calculation of eigenvalues, eigenvectors and other results needed for the execution of the canonical correlation (Johnson and Wichern, 1988). The following model was obtained for the case study:

$$0{,}5005 L_{eq} + 0{,}5802 D = 0{,}2260 X_1 + 0{,}1036 X_2 + 0{,}5681 X_3 + 0{,}3476 X_4 + $$
$$+ 0{,}4444 X_5 - 0{,}0502 X_6 + 0{,}0005 X_7$$

Where: X_1, X_2, X_3, X_4, X_5, X_6 and X_7 are variables of the questionnaire and the coefficients of the model are the canonical weights. For the variables of the subjective dataset one has:

X_1 – Classification of noise pollution as harmful to health;
X_2 – occurrence of discomfort due to noise pollution in the workplace;
X_3 – perception of noise in your workstation;
X_4 – perception of noise in neighboring workstations;
X_5 – reported physiological effects;
X_6 – factor of greatest disruption in the workplace;
X_7 – use of hearing protectors.

5. Interpretation of the results: The development of interpretations of the results is based on the statistical significance test of relationships between the two datasets and each of their indices (Hair et al., 1995). If there is significance, the interdependence between datasets is confirmed. The tests used were Wilks's lambda distribution, Pillai's criterion, Hotteling's trace and Roy's largest root (Hair et al., 1995, Johnson

and Wichern, 1988; Kendall, 1980). The tests indicated an association intensity of 75% between the two datasets.

6. Validation of the model: The model was tested on another sample to verify if it reacted as expected. For the case study, a sensitivity analysis was performed (Johnson and Wichern, 1988) which consisted of removing the variables X_1, X_2, X_6 and X_7 and performing a new calculation, which yielded by an association intensity of 73%.

The intergroup associations indicated the following:

1. Occurrence of doses above those established by the NR-15 standard at levels of around 82 dB(A) are not perceived by the exposed individual, especially in the nighttime work environment, which indicates a situation of risk that may lead the worker to not to use hearing protectors.
2. The physiological effects reported by employees are felt only after the exposure to noise at his workstation diminishes or stops. In addition to the fact that the worker does not feel an immediate physical reaction during the risk situation, this association may give the false impression that the physiological effect he feels is being caused only by the noise pollution from neighboring workstations that continue in operation.
3. The worker uses hearing protection only during the occurrence of high noise levels in his workstation, indicating that the issue of occupational safety is neglected during part of his workday.

CONCLUSION

Occupational noise in a service company of the public transport sector was evaluated. This evaluation involved the following steps: 1) Objective assessment based on the measurement and analysis of sound pressure levels and noise dose, measured in the workplace during day and night shifts; 2) Subjective assessment based on questionnaires to determine the workers' perception of noise in their work environment; and 3) A statistical analysis of the data obtained in steps 1 and 2.

Based on the analysis of the results of the objective assessment, the company's entire plant was found to present a situation of noise pollution that deserves attention, requiring the implementation of routine control actions, according to the classification of the NHO-01 standard. In addition to the noise pollution in their main workstation, the employees, through direct or indirect contact, are also affected by noise pollution from secondary workstations.

With regard to the parameters regulated by the NR 15 standard, the equivalent sound levels measured at workstations lay within the limits established for industrial noise, but the noise dose at various workstations in both day and night shifts was clearly above regulated levels, making it indispensable to use hearing protection.

Although the current standards do not consider subjective analysis a tool for judging findings, it is an important tool for characterizing and interpreting the problem of noise in working environments.

The results of the subjective assessment indicated that the workers had a low perception of noise levels generated in secondary workstations and an intermediate perception of noise in their own workstations, which is directly reflected in the low percentage of use of hearing protection. Although 100% of the respondents considered it important to use hearing protectors, only 75% of the respondents in the two shifts stated that they use them regularly, because they find these devices uncomfortable.

The canonical correlation indicated an association between the results of the objective evaluation and the subjective assessment, with a 75% association of intensity between the two datasets. Based on the associations between datasets, it was concluded that:

1. High noise doses at levels of about 82 dB(A) are not perceived by the exposed individual, resulting in his infrequent use of hearing protection.
2. Physiological effects are felt only after the exposure to noise in the workstation diminishes or ceases, causing the false impression that the physiological effects are being caused only by noise pollution from neighboring workstations that remain in operation.
3. Workers use hearing protection devices only during the occurrence of high levels of noise in their workstations, but neglect to use them during a large part of the workday.

ACKNOWLEDGMENTS

The authors gratefully acknowledge the German Government, through the German Academic Exchange Service - DAAD (Deutscher Akademischer Austauschdienst) and the Brazilian Government, through the National Council for Scientific and Technological Development - CNPq - for their financial support, which enabled the purchase of the sound level meters and software used in this study.

REFERENCES

Adera T, Donahue AM, Malit BD, Gaydos JC. Assessment of the proposed draft American National Standard Method for Evaluating the Effectiveness of Hearing Conservation Programs. *J Occup Med*, 1993, 35(6):568-573.

Alberti PW. Noise, the most ubiquitous pollutant. *Noise Health*, 1998, 1(1):3-5.

American National Standards Institute. ANSI S1.25: *Specification for Personal Noise Dosimeters*, Washington DC, 1991.

American National Standards Institute. ANSI S1.4: *Specification for Sound Level Meters*, Washington DC, 1983.

Associação Brasileira de Normas Técnicas [Brazilian Association of Technical Standards – ABNT]. NBR 10152: *Níveis de ruído para o conforto acústico* [Sound levels for acoustic comfort], Rio de Janeiro, 1987.

Barreto SM, Swerdlow AJ, Smith PG, Higgins CD. A nested case-control study of fatal work related injuries among Brazilian steel workers. *Occup Environ Med*, 1997, 54:599-604.

Berger EH, Royster LH, Royster JD, Driscoll DP, Layne M. *The noise manual*, Xxxx, 2000.

Berglund B, Lindvall T, Schwela DH. Guidelines for commuinity noise. *Published for the World Health Organization*, 1999.

Cordeiro R, Clemente AP, Diniz CS, Dias A. Occupational noise as a risk factor for work-related injuries. *Rev Saúde Pública*, 2005, 39(3):461-466.

Darlington P. Practical measurement of telecommunication receiver electroacoustics for the computation of acoustic dose. *In*: *Proceedings of the Institute of Acoustics*, St. Albans, 2003.

Dwyer T. *Life and death at work: industrial accidents as a case of socially produced error*. New York, 1991.

Fundação Jorge Duprat Figueiredo de Segurança e Medicina – Fundacentro. NHO 01: *Avaliação da exposição ocupacional ao ruído* [Standard NHO 01: Assessment of occupational noise exposure], São Paulo, 1999.

Hair JF, Anderson RE, Tatham RL, Black WC. *Multivariate Data Analysis*, New York, 1995.

International Electrotechnical Commission. IEC 651: *Sound Level Meters*, Geneva, 1979.

International Electrotechnical Commission. IEC 804: *Integrating-Averaging Sound Level Meters*, Geneva, 1985.

International Electrotechnical Commission. IEC 804: *Integrating-Averaging Sound Level Meters*, Geneva, 1985.

Johnson RA., Wichern DW. *Applied multivariate statistical analysis*, New Jersey, 1988.

Kendall M. *Multivariate analysis*, New York, 1980.

Kjellberg A. Subjective, behavioral and psychophysiological effects of noise. *Scand J Work Environ Health*, 1990, 16(1):29-38.

Melamed S, Luz J, Green MS. Noise exposure, noise annoyance and their relation to psychological distress, accident and sickness absence among blue-collar workers – the Cordis Study. *Isr J Med Sci*, 1992, 28(8-9):629-635.

Melamed S, Rabinowitz S, Green MS. Noise exposure, noise annoyance, use of hearing protection devices and distress among blue-collar workers. *Scand J Work Environ Health*, 1994, 20(4):294-300.

Ministério do Trabalho e Emprego [Brazilian Ministry of Labor and Employment]. NR 07: *Programa de controle médico de saúde ocupacional* [Standard NR 07: Program of Medical Control of Occupational Health], NR 09: *Programa de prevenção de riscos ambientais* [Standard NR 09: Environmental Risk Prevention Program], NR 15: *Atividades e Operações Insalubres* [Standard NR 15: Unhealthy Activities and Operations], NR 17: *Ergonomia* [Standard NR 17: Ergonomics]. Brasília, 1978.

Morata T, Dunn D. Occupational Medicine - State of The Art Reviews. *Occupational Hearing Loss*, 1995, 10(3): 641-656.

Occupational Safety and Health Administration - OSHA. *Occupational Noise Exposure: Hearing Conservation Amendment*, USA, 1981.

Paz EC, Ferreira AC, Zannin PHT. Estudo comparativo da percepção do ruído urbano [Comparative study of the perception of urban noise]. *Rev Saúde Pública*, 2005, 39: 467-472.

Portela BS, Zannin, PHT. Analysis of factors that influence noise levels inside urban buses. *Journal of Scientific & Industrial Research*, 2010, 69: 684-687.

Rothman KJ, Greenland S. *Modern Epidemiology*. Philadelphia, 1998.

World Health Organization. Regional Office for Europe. *Occupational and community noise*, 2003.

Zannin PHT. Occupational Noise in Urban Buses. *International Journal of Industrial Ergonomics*, 2006, 36(10): 901-905.

In: Noise and Ergonomics in the Workplace
Editor: Paulo Henrique Trombetta Zannin

ISBN: 978-1-62417-573-2

Chapter 13

NOISE IN URBAN BUSES: MEASUREMENTS AND BUS DRIVERS AND FARE COLLECTORS' PERCEPTION OF NOISE

Paulo Henrique Trombetta Zannin [*]
and Clifton Roberto Giovannini

Laboratory of Environmental and Industrial Acoustics and Acoustic Comfort (LAAICA),
Federal University of Paraná, PR, Brazil

ABSTRACT

This chapter discusses a study about noise perceived by urban bus drivers and fare collectors in the workplace. Interviews were held with 68 drivers and 32 fare collectors. Noise levels were measured in 60 buses. The noise levels in the bus drivers' workplace were evaluated according to: 1) ISO 1999 – Determination of occupational noise exposure and estimation of noise-induced hearing impairment; 2) *Occupational Hygiene Standards – Evaluation of Occupational Exposure to Noise*; and 3) the Brazilian Standard for Ergonomics. The statistical analysis indicated that the year of manufacture and the position of the engine are two factors that contribute significantly to the noise level in the workplace of bus drivers and fare collectors. Among the different bus categories evaluated, it can be concluded that feeder buses, whose engine is usually located at the front of the vehicle, generate the highest noise levels for drivers and fare collectors. The results of the interviews indicate that bus drivers and fare collectors consider the engine the main source of noise in buses. They perceive noise and are aware that it may impair their health. Both drivers and fare collectors claim that the main effects of noise they perceive are headaches, concentration difficulties, irritability, and buzzing in the ears. Among the interviewees, 21% of the drivers and 21% of the fare collectors reported having difficulties in understanding speech in the presence of noise, while 15% of the drivers and 24% of the fare collectors stated that audiometric testing revealed hearing loss.

[*] Email: paulo.zannin@pesquisador.cnpq.br; paulo.zannin@gmail.com

1. INTRODUCTION

The traffic noise generated by cars, buses, trains, trucks, and aircraft is a permanent (continuous or intermittent) source of noise in towns, small cities and especially large metropolises worldwide. Urban pollution, and particularly urban noise pollution, is an increasingly serious problem. Many surveys have been conducted to evaluate environmental noise in numerous cities and countries (Canelli, 1974; Brown, 1994; Chakrabarty et al., 1997; Zannin et al, 2002; Morillas et al., 2002; Korfali and Massoud, 2003; Zannin et al., 2003; Diniz and Zannin, 2004, Khilman, 2004; da Paz et al., 2005; Coensel et al., 2005; Zannin and Sant'Ana, 2011, Guedes et al., 2011). The impact of traffic noise on human health has also been studied worldwide (e.g., Griffiths and Langdon, 1986; Babish et al., 1999; Skinner and Grimwood, 2005).

Buses are an especially interesting object of study in terms of noise pollution in large cities. They are a source not only of urban environmental noise but also of workplace noise for drivers and fare collectors. Most Brazilian urban buses have a fare collector to collect passengers' fares. His workstation is usually located in the front part of the bus. Urban public bus transportation is an essential service rendered to society. Brazil, with its huge territory and its growing population, has a dense road network for urban, intermunicipal and interstate transportation.

The health and well-being of urban bus drivers has drawn the attention of epidemiologists, sociologists, physicians, psychologists and engineers (Evans and Johansson, 1998). Concerns about these workers are due to the fact that, according to Evans and Johansson (1998): *"the operation of buses in urban areas is a highly stressful and unhealthy professional activity"*. The work of bus drivers and fare collectors appears to be an especially stressful occupation due to the diversity of possible physical and/or psychosocial stressors (Guidotti and Cottle, 1987; Evans and Carrère, 1991; Rydstedt et.al., 1998). Physical stressors include, for example, traffic congestion, safety-related problems due to possible traffic accidents, ergonomic problems, fluctuations in temperature with opening and closing of doors, whole body vibration, hand-arm vibration, noise emitted by the engine, noise emitted by urban traffic around the bus, noise generated by opening and closing of doors, etc. (Guidotti and Cottle, 1987; Evans and Carrère, 1991; Rydstedt et.al., 1998, Portela and Zannin, 2010). Another constant stressor for drivers and fare collectors is the concern for safety in the event of robbery.

Fonseca (1993) evaluated the occupational exposure to noise of urban bus drivers and fare collectors in São Paulo, one of the largest cities in the world. He concluded that the measured noise dose is related directly with the location of the engine. It was also found that measured noises doses in buses whose engines are located in the front exceed or are very close to the limit value of this dose, D = 100%, which is established by national standards. He evaluated 12 buses, 7 of which had the engine located in the front. Of these 7 buses, 5 exceeded the dose of D = 100%. It was also found that none of the doses measured for the fare collectors exceeded 100%.

Latance Júnior (1999) evaluated 40 different buses, none of which showed a dose above D = 100%. The measurements were taken with a noise dosimeter, and the evaluation covered the bus drivers' entire workday.

Becker and Tamagna (1998) studied noise in urban buses in the city of Porto Alegre, Brazil, analyzing noise internal and external to buses. All the internal noise measurements, with a sampling time of 1 minute, were taken along the same route, but without passengers. They concluded that the main source of noise is the engine. Their results indicated that the measured noise levels in the newer vehicles were significantly lower than in the older ones, especially in the positions closer to the engine. Their internal measurements showed no equivalent sound level, L_{eq}, value that reached 85 dB(A). Considering that according to Brazilian regulations, the level of L_{eq} = 85 dB(A) associated to an exposure time of 8 hours results in a dose of 100%, none of the measured levels reached this value.

Tse et al. (2006) make an extensive review of the last 50 years of research on the theme of "Bus driver well-being." However, despite the excellence and vastness of their review, it does not offer any comments about the problem of noise in the bus driver's workplace.

Curitiba's public transportation system has been the recipient of several international awards due to its efficiency. One award was bestowed by the Building and Social Housing Foundation, and another by the Worldwatch Institute, which described Curitiba's urban transportation system as an exemplary one for other large cities. Curitiba's public transportation system has also been the subject of several studies (Herbst, 1992; Smith and Hensher, 1998; Diesendorf, 2000). Herbst (1992) stated that: "*Curitiba today probably has the world's most efficient city bus system*". Smith and Hensher (1998) stated that "*it is an example of effective land use and transport planning, and creative and efficient provision of transport services*".

Curitiba's Integrated Transport Network system comprises 2,160 buses that operate 469 different lines across the city. More than two million passengers use buses in Curitiba each workday (Zannin, 2006; Zannin, 2008).

Despite the efficiency and modernity of today's public transport systems, however, the responsibility for driving falls upon a single person, the driver. According to Gulian et al. (1989), "*factors such as general life conditions (origin, level of schooling, housing, food), general working conditions (medical assistance, characteristics of the bus, duration of the workday), workplace environment (noise, vibration, high temperature, chemical pollutants) and other factors will trigger adverse situations in the health of these workers, which may frequently affect them negatively*".

In view of the complexity of the theme, as described above, the purpose of this chapter is to evaluate occupational exposure to and perception of noise in the workplace of urban bus drivers and fare collectors. To this end, two surveys, one objective and the other subjective, were conducted as follows:

1) Objective evaluation: Characterize the workplace environment by measuring sound pressure levels, and calculating the noise dose and the standardized exposure level.
2) Subjective assessment: Based on the answers to a questionnaire, characterize the workers' perception of noise in their workplace environment.

2. Evaluation Metrics for Workplace Noise

In Brazil, the metric for assessing occupational noise is Standard NR15 of the Brazilian Ministry of Labor and Employment. This standard, which was promulgated into law in 1978, is outdated and out of step with international standards of occupational noise assessment, such as those established by ISO 1999 – Acoustics – Determination of occupational noise exposure and estimation of noise-induced hearing impairment (1990). On the other hand, Brazil has an Occupational Hygiene Standard – NHO-01 (Fundacentro, 1999), which describes a technical procedure for the Evaluation of Occupational Exposure to Noise, which follows the guidelines of ISO 1999 (1990). The NHO-01 standard was developed by occupational hygiene specialists of the Jorge Duprat Figueiredo Foundation for Occupational Hygiene and Safety – Fundacentro.

According to the NHO-01 standard, occupational exposure to continuous or intermittent noise is evaluated by determining the daily noise dose, D, or the exposure level, L_E. The NHO-01 standard establishes an exposure time T_E = 8 hours (or 480 minutes) and the exposure level L_E = 85 dB(A) as reference criteria for the evaluation of occupational exposure. These T_E and L_E values correspond to the noise dose limit of D = 100% for a working day.

The daily noise dose and exposure level can be determined by means of the following expressions (Fundacentro, 1999):

$$L_E = 10\log\left((480/T_E).(D/100)\right) + 85 \text{ [dB(A)]} \quad (1)$$

$$D = (T_E/480).100.2^{((L_E - 85)/3)} \quad [\%] \quad (2)$$

where:

- L_E is the daily noise exposure level expressed in dB and A-weighted. In practice, L_E corresponds to the equivalent sound level L_{Aeq}, which is measured during the working day considering the exposure time T_E;
- T_E is the exposure time during the working day, and is expressed in [hours or minutes];
- D is the noise dose received daily by the worker, and is expressed in [%].

The fundamental difference between the NR15 and NHO-01 standards is illustrated in Equation [2]. This difference stems from the fact that the NR15 standard uses the Exchange rate q = 5 dB(A) as the noise D dose doubling factor while the NHO-01 standard uses the Exchange rate q = 3 dB(A). To exemplify, let us consider an exposure level L_E = 90 dB(A) and T_E = 8 h. Considering that the daily noise dose limit is D = 100%, which should not be exceeded, then for q = 5 and L_E = 90 dB(A) one has a maximum permissible exposure time of T_E = 4 hours. Considering the Exchange rate q = 3 and L_E = 90 dB(A), the maximum permissible exposure would be T_E = 2 hours, 31 minutes and 43 seconds. Hence, the use of q = 5 may place the worker in a situation of underevaluation of occupational noise.

Table 1, adapted from the book Occupational Exposure to Noise: Evaluation, Prevention and Control (Goelzer et al., 2001) and published under the auspices of the World Health Organization – WHO, presents a few examples of parameters for the evaluation of occupational noise used in several countries.

Table 1. Parameters for the evaluation of occupational noise

Country	8-hour average sound pressure level dB(A)	Exchange rate "q" dB(A)
Argentina	90	3
Australia	85	3
Brazil	85	5
Chile	85	5
France	85	3
Germany	85	3
Israel	85	5
Italy	85	3
Netherlands	85	3
New Zealand	85	3
Norway	85	3
USA (Army and Air Force)	85	3
Uruguay	90	3

Adapted from Goelzer et al., 2001.

Table 1 shows that there is no worldwide consensus about the use of the Exchange rate "q," i.e., some countries adopt q = 3 and others q = 5. On the other hand, most of the countries listed in Table 1 adopt the Exchange rate q = 3. Goelzer et al.'s book (2001), which is a special report sponsored by WHO, recommends the parameters listed in Table 2:

Table 2. Parameters recommended by the WHO for the evaluation of occupational noise

Occupational Exposure to Noise: Evaluation, Prevention and Control - WHO -	Noise exposure level standardized to a nominal 8 hour working day $L_{EX,8h}$ dB(A)	Exchange rate "q" dB(A)
Special Report – WHO S64	85	3

Adapted from Goelzer et al., 2001.

The NHO-01 standard, which follows the guidelines of the ISO 1999 standard, establishes as the criterion of occupational exposure to noise during a working day, the parameter – noise exposure level standardized to a nominal 8-hour working day – $L_{EX,8h}$. This parameter is given by equation:

$$L_{EX,8h} = L_{eq,TE} + 10\log (T_E/T_0) \text{ [dB(A)]} \quad (3)$$

where:

- T_E is the exposure time, i.e., the effective duration of the working day, and
- T_0 is the reference duration of the working day. T_0 = 480 minutes or 8 hours.

According to Goelzer et al. (2001): "*the term daily personal noise exposure of a worker* $L_{EP,d}$ is the same as the term *noise exposure level standardized to a standard 8-h working day*, $L_{EX,8h}$." Both terms are expressed in dB(A).

For the purpose of determining noise exposure, the Occupational Hygiene Standard NHO-01 considers that the value of reference for the standardized exposure level, $L_{EX,8h}$, should be equal to the maximum of 85 dB(A), considering an 8-hour working day. Values exceeding $L_{EX,8h}$ = 85 dB(A) indicate occupational noise pollution and measures should be taken to reduce these noise levels.

As the criterion to determine and make decisions concerning occupational noise problems, the NHO-01 standard recommends the procedures described in Table 3:

Table 3. Determination of occupational noise levels and decision-making criteria

Daily dose D (%)	**Standardized Exposure Level $L_{EX,8h}$ dB(A)**	**Technical consideration**	**Required action**
0 – 50	Below 82	Acceptable	At least the maintenance of the current condition
50 – 80	82 – 84	Action level	Adoption of preventive measures
80 – 100	84 – 85	Uncertainty region	Adoption of preventive and corrective measures to seek the daily dose reduction
Above 100	Above 85	Above exposure level	Immediate adoption of corrective measures

In the case of Brazil, Ergonomic Standard NR-17 of the Ministry of Labor and Employment (2011) can be used to evaluate the conditions of the workplace environment together with the workers' psychophysiological characteristics. This ergonomic standard establishes assessment parameters that are indicators of comfort, safety and performance in the workplace environment.

The ergonomic standard NR-17 establishes the following characteristics for workplace environments: "*In workplaces where activities that require constant attention and intellectual effort are performed, such as control rooms, laboratories*, project analysis or development *offices, among others, the following conditions of comfort are recommended*":

1) An effective temperature index of 20 to 23°C,
2) Airspeed not exceeding 0.75 m/s,
3) Relative humidity of no less than 40%,
4) Noise levels as established by the Brazilian NBR 10152 standard – Sound levels for acoustic comfort in the built environment.
   4.1) For activities with the above defined characteristics but that are not equivalent to or correlated with those listed in the NBR 10152 standard, the acceptable noise level for comfort shall be up to 65 dB (A).

Although the occupation of bus driver does not appear to have the characteristics cited in the NR-17 standard, this occupation is obviously stressful and requires constant attention in an urban environment increasingly congested with vehicles and people. The current literature identifies the characteristics peculiar to the occupation of bus driver, as can be seen in Evans and Johansson (1998): "*Urban bus operation appears to be an especially stressful occupation.*" Guidotti and Cottle (1987) and Evans and Carrère (1991) describe several stress factors to which bus drivers are subjected, i.e., traffic congestion, ergonomic problems, fluctuations in temperature with opening and closing of doors, vibration and noise. Portella and Zannin (2010) evaluated noise in urban buses and concluded that "*a comparison of older [bus] models against newer ones indicated that the workplace environment in older buses was more unfavorable for drivers' performance.*"

In view of the above, this chapter also evaluates the bus driver's workplace environment using the ergonomic standard NR-17. The parameter of analysis, the equivalent sound pressure level L_{eq}, should be at the most 65 dB(A) for the workplace environment to be considered comfortable. Noise levels above 65 dB(A) are therefore considered uncomfortable.

3. METHODOLOGY

3.1. Description of the Buses

Noise levels were evaluated in 60 urban buses. Three models of the urban fleet, which represent the majority of the buses in circulation, were chosen for analysis: 1) Feeder buses, 2) Speedy buses, and 3) Bi-articulated buses. Twenty units of each model were evaluated.

Feeder buses are ordinary buses with a door at the front for passengers to get on the bus, and are operated by two workers, a driver and a fare collector. These vehicles follow pre-established routes using the same traffic lanes as passenger cars, motorbikes, utility vehicles, etc. Feeder buses operate normally throughout the city, along regular streets, picking up and dropping off passengers at conventional bus stops, as indicated in Figure 1.

Figure 1. Feeder bus.

Buses operating on direct line runs, called speedy buses, do not have fare collectors inside because the latter work in the tube-shaped bus stops destined for passenger *boarding and de-boarding* (see Figure 2a and 2b). These buses travel on ordinary streets, but the distance between stops is much greater than that the feeder or bi-articulated buses, since they stop only at tube-shaped bus stops, which are fewer (see Figure 2a and 2b).

(A)

(B)

Figure 2. (A) Speedy bus. (B) Tube-shaped bus stop for Speedy bus.

The bi-articulated buses travel only along special channels, which are exclusively for this type of bus, and stop at their tube stops. No cars, motorcycles, or other vehicles are allowed to travel in these channels or exclusive bus corridors. Bi-articulated buses are much larger than speedy buses because they have two hinges and three blocks, while speedy buses are single block vehicles. The fare collectors for this bus system work in the tube stops instead of in the buses. The tube stops are special stops for rapid passenger *boarding and de-boarding* (see Figures 2b and 3).

Figure 3. Bi-articulated bus.

3.2. Objective evaluation – Measurements

Measurements were taken on several bus routes under standard operating conditions, i.e., buses occupied by passengers (Zannin, 2006; Zannin, 2008). The measurements were taken only in sunny weather, on weekdays, and at times not coinciding with the rush hour. No measurements were taken on weekends or holidays. Depending on the type of bus and distance of the route, measurements were taken along the entire route from the time a bus left a particular stop until it returned to the same stop. In some cases, selected several stops were selected along the route so that the researchers could exit the bus they were traveling in and board the next one, optimizing the measurement time and the number of buses evaluated.

The measurements were taken according to the ISO 1999 standard – "Determination of occupational noise exposure and estimate of noise-induced hearing impairment" (1990), the Occupational Hygiene Standard NHO-01 (1999) and the Brazilian Standard for Ergonomics NR-17 (2011). The microphone of the sound level meter was placed at 0.10 ± 0.01 m from the left ear of the bus driver (ISO, 1990). The ear assessed for incoming noise was the one receiving the higher value of the equivalent continuous A-weighted sound pressure level $L_{eq,T}$ in dB(A). The noise exposure level was standardized to a nominal 8-h working day $L_{EX,8h}$ calculated from the measured equivalent sound pressure level $L_{eq,T}$. Standardized exposure levels were calculated using Protector 7825 software (Brüel and Kjær, 1999):

$$L_{EX,8h} = L_{eq,T_e} + 10\lg\frac{T_e}{T_0} \quad \text{dB(A)} \tag{4}$$

where:

L_{eq,T_e} is the equivalent continuous A-weighted sound pressure level

T_e is the effective duration of the working day

T_0 is the reference duration of the working day (8 h or 480 minutes).

The workday of bus drivers in Curitiba is 6 hours, with a brief 10 - 15 min break after 3 h of driving. Therefore, the exposure time considered in equation [4] was T_e = 6 h. The 6-h workday is the norm for Curitiba bus drivers (URBS, 2012).

The workplaces were evaluated based on the Occupational Hygiene Standard NHO-01 (1999). Table 3 describes the parameters for the determination of occupational noise levels and the decision-making criteria according to the NHO-01 standard, considering the calculated values of standardized exposure level $L_{EX,8h}$ and a daily noise D dose (Zannin, 2006, Zannin 2008).

The Brazilian Standard for Ergonomics NR-17 (2011) was also considered in the evaluation of noise levels inside the buses. With respect to noise immission, this Standard classifies a workplace as comfortable or uncomfortable. For a workplace to be considered comfortable, the equivalent sound pressure level L_{eq} should not exceed or be at the most equal to 65 dB(A). Therefore, workplaces where L_{eq} exceeds 65 dB(A) are considered uncomfortable.

The measurements were taken with Brüel and Kjær Mediator 2238 type 1 and the Brüel and Kjær Investigator 2260 type 1 integrating and logging sound level meters (Brüel and Kjær, 1998). Both instruments were set in the fast response mode, using the A-weighing curve. The sound level meters were calibrated regularly in the laboratory, before beginning each measuring cycle, and at the end of each fieldwork session, using a BK 4231 sound calibrator (Brüel and Kjær, 1998). All the standards used here as reference allow for the use of sound pressure level meters such as those used in this study (Zannin, 2006; Zannin, 2008).

Table 4. Bi-articulated buses

Measurement no.	L_{eq} dB(A)	$L_{EX,8h}$ dB(A)	Dose [%]	Year of manufacture	Location of the engine
1	80.2	79	24.8	1995	M
2	79.6	78.4	21.6	1995	M
3	79.2	78	19.7	1995	M
4	78.8	77.6	18	1995	M
5	78.7	77.5	17.6	1995	M
6	78.7	77.5	17.6	1995	M
7	78.5	77.4	17.2	1995	M
8	78	76.8	15	1997	M
9	78	76.8	15	1995	M
10	77.9	76.7	14.6	1995	M
11	77.2	76	12.4	1995	M
12	77	75.8	11.9	1997	M
13	76.9	75.7	11.6	1995	M
14	76.8	75.6	11.4	1995	M
15	76.8	75.6	11.4	1995	M
16	76.8	75.6	11.4	1995	M
17	76.3	75.1	10.1	1995	M
18	75.7	74.5	8.8	1995	M
19	75.1	73.9	7.7	2000	M
20	75	73.8	7.5	2000	M

3.3. Subjective Assessment – Questionnaires

In this part of the study on occupational noise in urban buses, a questionnaire was used to identify the bus drivers and fare collectors' reactions to perceived noise. Some of the questions were designed to be answered based on a scale. After a pilot study, it was decided to eliminate the scale, leaving only questions to be answered with a check mark, since it was found that the respondents found it very difficult to use the scale. This difficulty may be due to impatience or tiredness, since the interviews always were conducted at the end of a workday. One hundred interviews were held, 68 involving drivers and 32 involving fare collectors. Each interviewee received a questionnaire, filled it out and returned it to the interviewer. Basically, the questionnaire sought to evaluate the following factors: a) Age, sex, employment period, b) If the worker is bothered by noise or not, c) If the worker thinks the noise generated by noise sources should be reduced, d) Which noise sources generate discomfort, and e) Symptoms the worker perceives as physical effects resulting from noise.

4. RESULTS AND DISCUSSION

Tables 4, 5 and 6 describe the type of bus, year of manufacture, location of the engine, equivalent sound pressure level L_{eq}, standardized exposure levels $L_{EX,8h}$ and daily noise D dose. The following parameters were used for the calculations: Exchange rate q =3, exposure time T_E = 6 hours, and standard time T_0 = 8 hours. In these tables, the location of the engine is defined as follows: F = Front, M = Middle-mounted in first car, R = Rear.

Table 5. Speedy buses

Measurement no.	L_{eq} dB(A)	$L_{EX,8h}$ dB(A)	Dose [%]	Year of manufacture	Location of the engine
1	77.8	78.2	20.7	1999	M
2	73.1	76.6	14.3	1996	M
3	77.2	76.0	12.4	1997	M
4	77	75.8	11.9	2000	M
5	76.8	75.6	11.4	1997	M
6	75.6	74.4	8.6	2000	R
7	75.6	74.4	8.6	2000	R
8	75.4	74.2	8.2	2002	M
9	74	72.8	6.0	2000	R
10	74	72.8	6.0	2001	R
11	73.7	72.5	5.6	2001	R
12	73.4	72.2	5.2	2000	R
13	73.3	72.1	5.1	2001	R
14	73.1	71.9	4.8	2001	R
15	73.0	71.8	4.7	2000	R
16	72.3	71.1	4	1999	R
17	71.8	70.6	3.6	2000	R
18	71.6	70.4	3.4	2000	R
19	70.3	69.1	2.5	2000	R
20	69.1	67.9	1.9	2000	R

The data in Tables 4, 5, and 6 indicate that the bi-articulated buses, which only have middle-mounted engines, exposed their drivers to noise levels of $73.8 \leq L_{EX,8h} \leq 79$ dB(A). Speedy buses with rear- and middle-mounted engines exposed drivers to noise levels of

$67.9 \leq L_{EX,8h} \leq 74.4$ dB(A). Feeder buses with rear-mounted engines exposed drivers to noise levels of $74.7 \leq L_{EX,8h} \leq 76.9$ dB(A) (Zannin, 2006; Zannin, 2008).

Table 6. Feeder buses

Measurement no.	L_{eq} dB(A)	$L_{EX,8h}$ dB(A)	Dose [%]	Year of manufacture	Location of the engine
1	86.7	**85.5**	**110.9**	1994	F
2	84.9	**83.7**	**73.3**	1991	F
3	84.2	**83**	**62.4**	1992	F
4	83.5	**82.3**	**53.1**	1991	F
5	83.3	**82.1**	**50.7**	1998	F
6	82.7	81.5	44.2	1998	F
7	82	80.8	37.6	1997	F
8	81.7	80.5	35.1	1998	F
9	81.7	80.5	35.1	1996	F
10	81.2	80	31.3	1998	F
11	81.2	80	31.3	1998	F
12	81.1	79.9	30.6	1999	F
13	81.8	79.6	28.5	1998	F
14	80.3	79.1	25.4	1998	F
15	80.1	78.9	24.3	1998	F
16	79.9	78.6	23.1	1998	F
17	78.1	76.9	15.3	1992	R
18	77.4	76.2	13	1992	R
19	76.3	75.1	10.1	1992	R
20	75.9	74.7	9.2	1992	R

Tables 4, 5 and 6 indicated that only five buses presented D dose higher than 50% and an exposure level $L_{EX,8h}$ above 82 dB(A). Only one bus showed D dose exceeding 100% and an exposure level $L_{EX,8h}$ above 85 dB(A). According to Table 3, in situations in which the D dose ranges from 50 to 80% and the exposure level $L_{EX,8h}$ from 82 to 84 dB(A), preventive measures should be adopted to reduce the noise level. In situations in which the D dose exceeds 100% and the exposure level $L_{EX,8h}$ exceeds 85 dB(A), which was the case of a single bus, noise reduction measures should be adopted immediately.

The Occupational Hygiene Standard and the WHO Special Report on Occupational Noise (Goelzer et al., 2001) adopt the standardized exposure level $L_{EX,8h}$ as the parameter of evaluation, with an occupational noise exposure limit of $L_{EX,8h} = 85$ dB(A). According to this criterion, therefore, the workplace for bus drivers can be considered "safe," except in one bus, which presented $L_{EX,8h} > 85$ dB(A) (see Tables 4, 5, and 6). As for the daily noise dose, the drivers' workplace can also be considered "safe," since a D dose exceeding 100%, i.e., 100.9%, was found in only one bus.

When the ergonomic standard NR-17 is adopted to evaluate the workplace environment of bus drivers, the situation differs from the one described above. This standard considers the workplace comfortable if the equivalent continuous sound pressure level L_{eq} is at most 65 dB(A). An analysis of Tables 4, 5 and 6 clearly indicates that none of the evaluated buses meets this condition. All the buses were rated uncomfortable, since the L_{eq} exceeded 65 dB(A). The situation was worse in relatively older buses and buses with front-mounted engines.

Figure 4 illustrates all the evaluated bus models and the equivalent sound pressure levels they generate. Note that the feeder buses are the noisiest ones and the speedy buses the most silent.

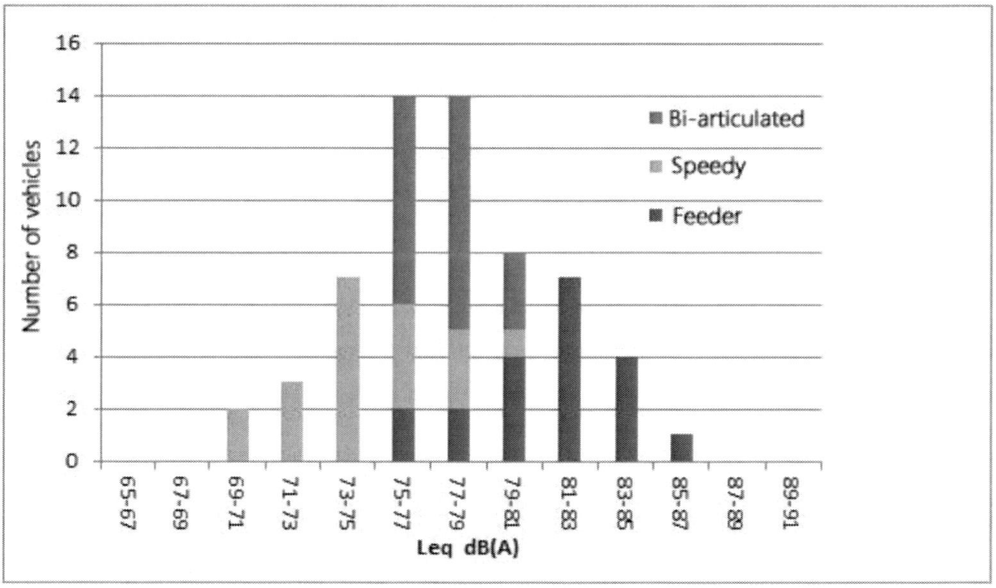

Figure 4. Number of buses evaluated as a function of the equivalent sound pressure level L_{eq}.

Figure 5 below, which was plotted from the data listed in Tables 4, 5 and 6 (Zannin, 2006; 2008), shows significant negative correlations between the year of manufacture and measured noise levels, considering jointly the three engine locations and all the bus types. Note the variability in noise levels measured in buses manufactured in the same year, as also illustrated in Figures 5 and 6.

As Figure 7 shows, in all the types of buses, the mean noise levels for the driver are higher in buses with engines located in the front, lower when located in the middle, and still lower when located in the rear. Therefore, drivers working in buses with front-mounted engines are at higher risk of hearing loss or work-related health problems.

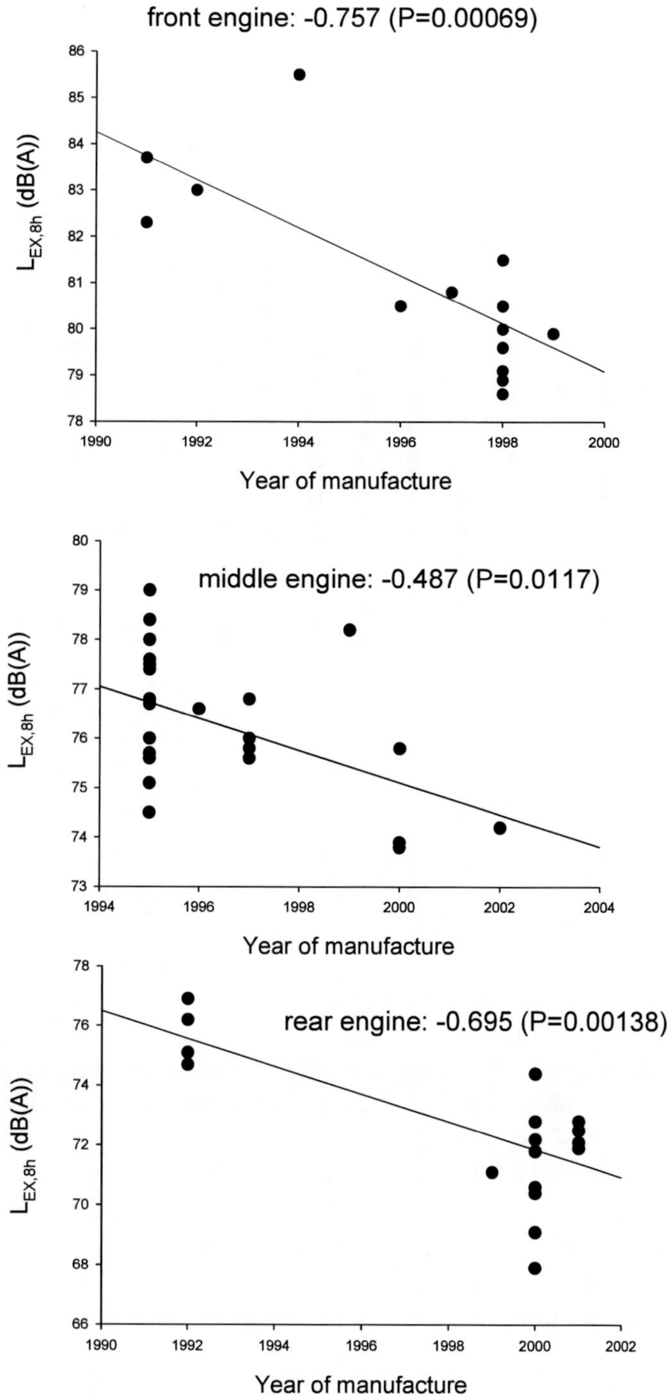

Original data taken from Zannin (2006; 2008).

Figure 5. Daily noise exposure ($L_{Ex,8h}$ dB(A)) and its correlation with the year of bus manufacture in buses with front, middle or rear-mounted engines. The location of the engine is followed by the Pearson correlation coefficient, with the P value in parenthesis.

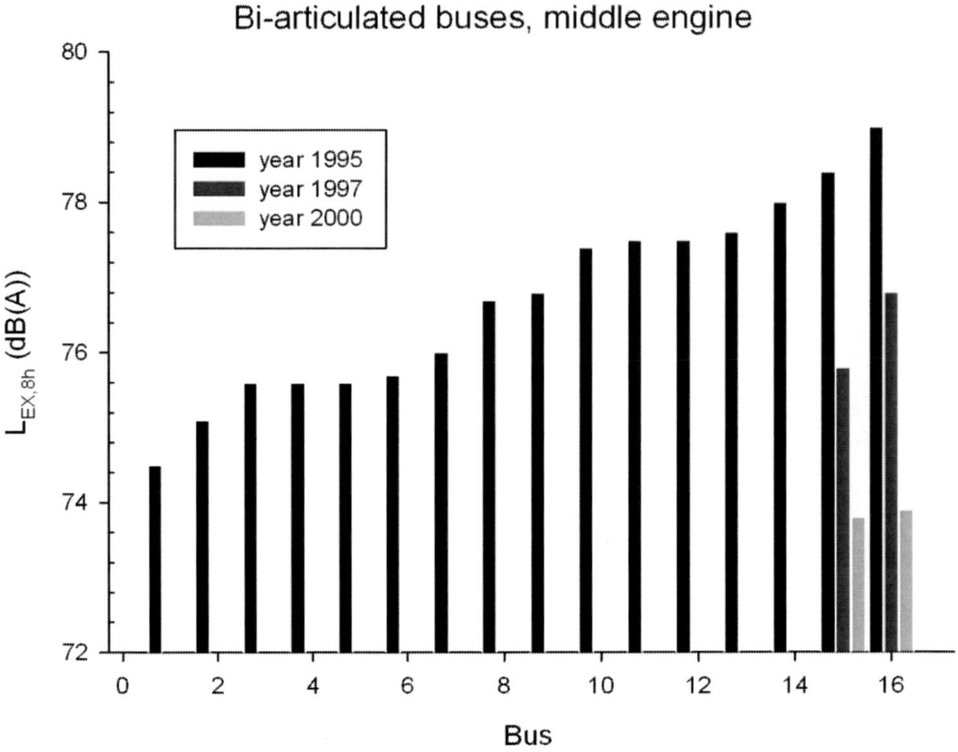

Figure 6. Daily noise exposure ($L_{Ex,8h}$ dB(A)) in bi-articulated buses with middle-mounted engines.

All the interviews with drivers and fare collectors were held at the end of their shift. The assessment involved 68 drivers and 32 fare collectors, all men, ranging in age from 35 to 57.

When asked if they believe that the noise in their workplace environment may be harmful to their health, and particularly to their hearing, 74% of the drivers answered affirmatively, while 26% answered in the negative. Among the fare collectors 88% stated they think it does, and 12% stated they did not think so. With regard to the need to reduce the noise level inside buses, 80% of the drivers stated that such a reduction was necessary and 20% stated it was not. Among the fare collectors, 91% considered noise reduction inside buses necessary and only 9% considered this measure unnecessary.

Figure 8 illustrates the period of time the interviewees have been working as urban bus drivers. As can be seen, a large majority of drivers, i.e., 65%, have been working in this profession for more than five years, and 33% have been in the profession for over ten years. Less than 10% of the sample, i.e. 7%, have worked as bus drivers for only one year.

Figure 9 indicates that, unlike the drivers, the majority of bus fare collectors, or 56% of the total sample, have worked in this profession for at most 5 years. Only 8% of the fare collectors have worked in this profession for more than 10 years. This is explained by that fact that fare collectors are usually younger individuals whose occupation requires less experience than that of bus drivers.

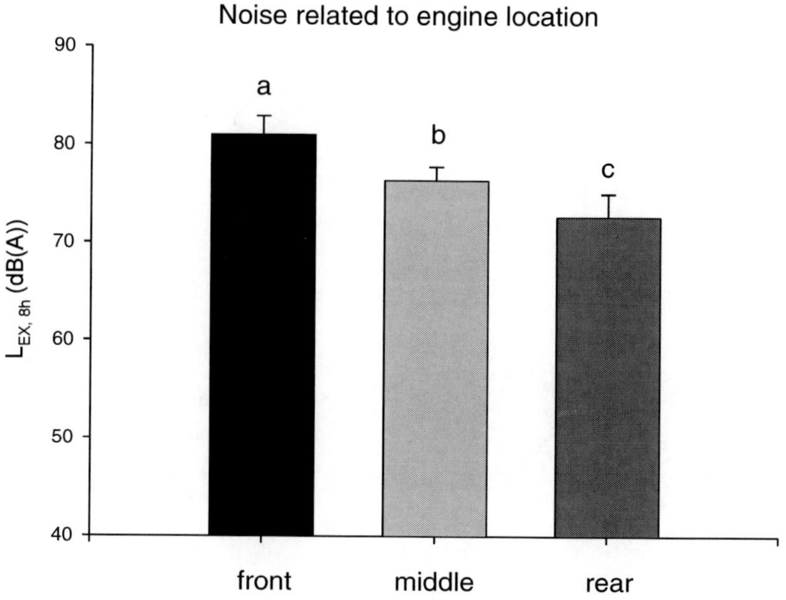

Figure 7. Daily noise exposure ($L_{Ex,8h}$ dB(A)) (mean ± SD) as a function of engine location in the three types of buses examined by Zannin (2006, 2008): bi-articulated, speedy, and feeders. Data from Tables 4-6, total n = 60 buses. One-way ANOVA followed by *post hoc* Tukey test indicated that the 3 groups differ from one another ($p < 0.05$).

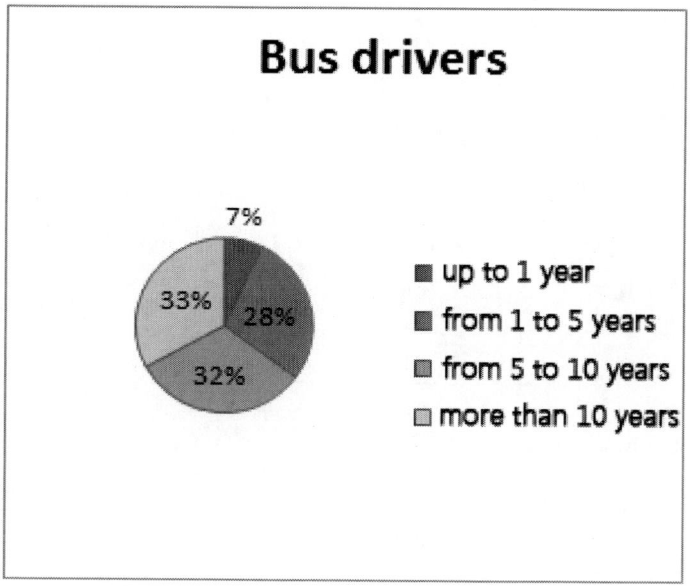

Figure 8. Answers to the question: "How long have you worked in this profession?"

The drivers were also asked if they always work in the same bus, to which 70% answered no, while 30% answered yes. This finding, allied to the period of time working as a driver, as shown in Figure 4, indicates that many drivers are familiar with the characteristics of their workplace environment, buses, and can perceive which of them, or which types of buses are noisier, and are also able to identify the sources of noise inside the buses. Among the fare collectors, 76% stated that they do not always work in the same bus, while 24% stated that they do.

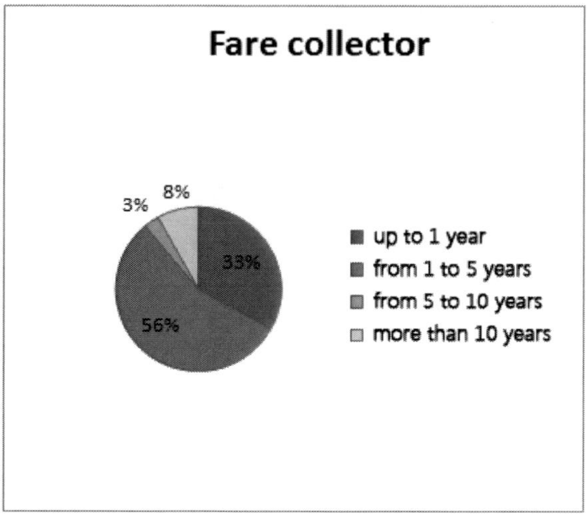

Figure 9. Answers to the question: "How long have you worked as a fare collector in buses?

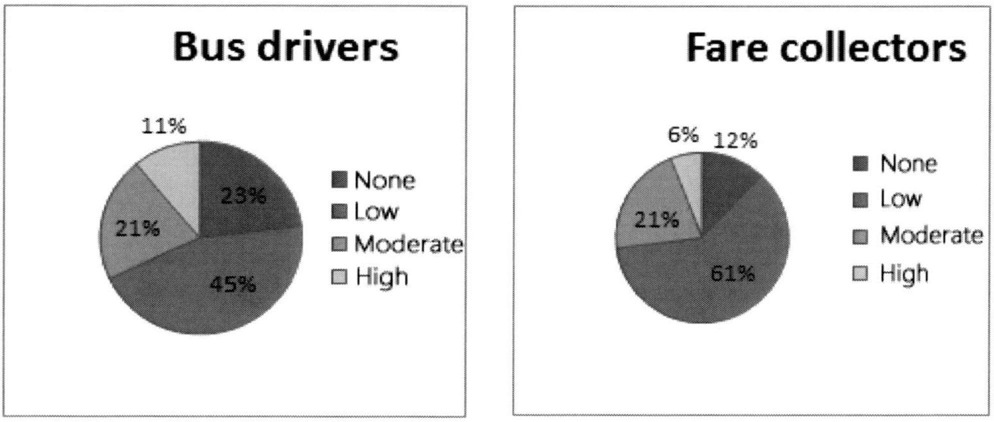

Figure 10. Degree of noise-induced discomfort in buses.

Drivers and fare collectors were asked to evaluate the degree of discomfort they perceive during their daily activities in buses. Figure 10 illustrates their responses.

Figure 10 indicates that most of the workers, i.e., 77% of the drivers and 88% of the fare collectors, feel irked (low, moderate and high degree of annoyance) by the noise generated inside buses, since only 23% of the drivers and 12% of the fare collectors stated they feel no discomfort.

These findings are in line with the measured noise levels, confirming that they are very high for a work environment that requires constant concentration. According to the analysis of the drivers' workplace environment using the ergonomic standard NR-17, all the buses can be considered uncomfortable, since the measured noise level was higher than 65 dB(A).

Drivers and fare collectors were also asked to indicate the main sources of noise they perceive during their workday and which cause discomfort. Figure 11 illustrates the main sources of discomfort.

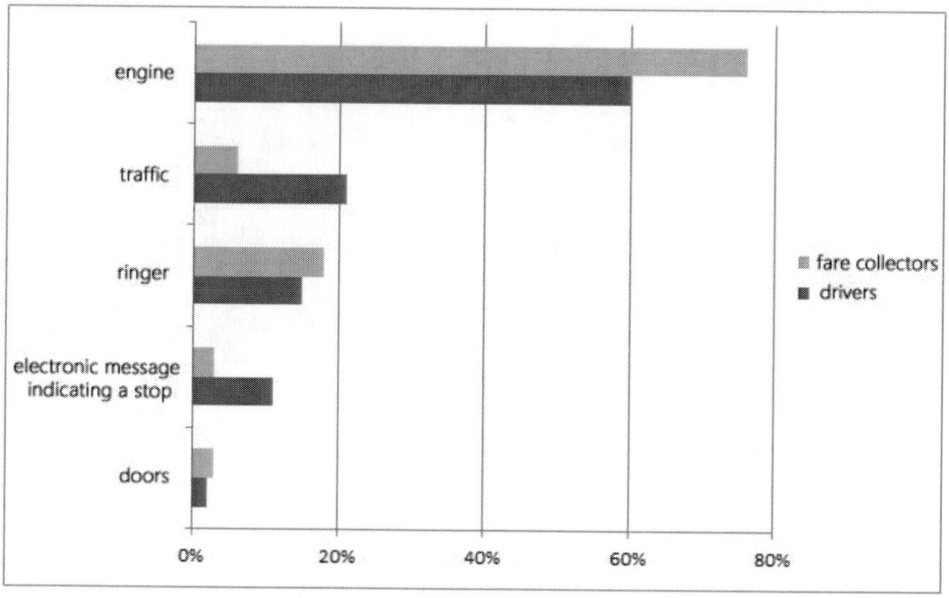

Figure 11. Main sources of discomfort during the workday.

The drivers and fare collectors' answers indicate that the engine is the main source of noise. The engine has also been identified in other studies as the main source of noise (Fonseca, 1993; Becker and Tamagna, 1998; Latance Júnior, 1999; Portella and Zannin, 2010). With regard to this noise source, the engine's noise emissions could be reduced in the engine design phase. Mounting the engine at the rear of the bus also reduces the noise level for the driver, as indicated in Figure 7, which correlates the location of the engine with noise levels. Noise emission levels produced by the running engine could be reduced by enclosing the engine.

Figure 12 shows the noise spectrum generated by the engine of a bi-articulated bus, measured in the frequency range of 12.5 Hz to 20000 Hz. This measurement was taken at the driver's workstation while the engine was revved. Figure 12 illustrates the equivalent sound pressure levels, and the maximum and minimum sound levels. The total equivalent sound level resulting from the sum of the parts in each frequency band reached a value of 79.6 dB(A), while the total maximum sound level showed a value of L_{max} = 84.5 dB(A) and the total minimum sound level a value of L_{min} = 70 dB(A). The equivalent sound level between 250 Hz and 2500 Hz ranged from 66.8 to 65.6 dB(A), i.e., it exceeded the level of comfort of L_{eq} = 65 dB(A), according to the Brazilian Standard for Ergonomics NR-17.

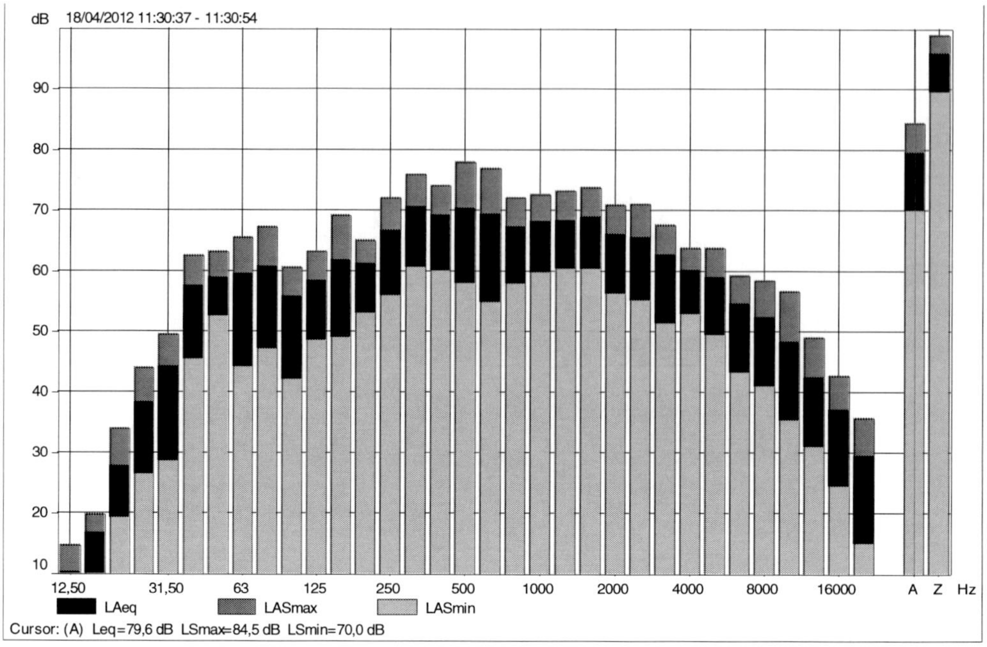

Figure 12. Noise spectrum generated by the engine of a bi-articulated bus.

With regard to the problem of traffic noise, which is considered the second highest source of discomfort for drivers and the third highest source for fare collectors, this problem is difficult to solve because it involves a variety of actors in the scenario of environmental pollution. The large Brazilian cities, and Curitiba in particular, present high environmental noise levels, and the population attributes the main source of noise to vehicle traffic (Zannin et al. 2002, Zannin et al., 2003; da Paz et al., 2005).

The third main source of discomfort for drivers and the second one for fare collectors is the ringer. This device is activated by passengers to indicate they want to de-board at a given bus stop, and may be activated several times for the same stop. The ringer is located in the front part of the bus next to the driver and fare collector's workstations, which is the reason they indicated it as a relevant source of discomfort. The sound spectrum of the ringer is illustrated in Figure 13. The measurements of the sound spectrum of the ringer were taken in stationary buses with their engines turned off. As this figure shows, the predominant sound component of this device is of high frequency, more specifically the frequency of 4000 Hz, which makes the sound it emits extremely annoying. The overall sound level at 4000 Hz is high, reaching up to $L_{eq} = 73$ dB(A).

Figure 14 indicates how noise strongly affects the physical well-being of drivers during their work shift, causing, in particular: headaches, 41%, irritability, 32%, difficulty to concentrate, 27% and buzzing in the ears, 24%. The symptoms perceived by fare collectors are: irritability, 39%; difficulty to concentrate, 36%; headaches, 30%; and buzzing in the ears, 12%. All these symptoms, individually or combined, may impair the physical health of these workers. In addition, they may also increase the potential for accidents, especially traffic accidents, since driving a bus requires constant attention, which may be impaired by headaches, irritability, concentration difficulties, etc., as pointed out by the interviewed drivers.

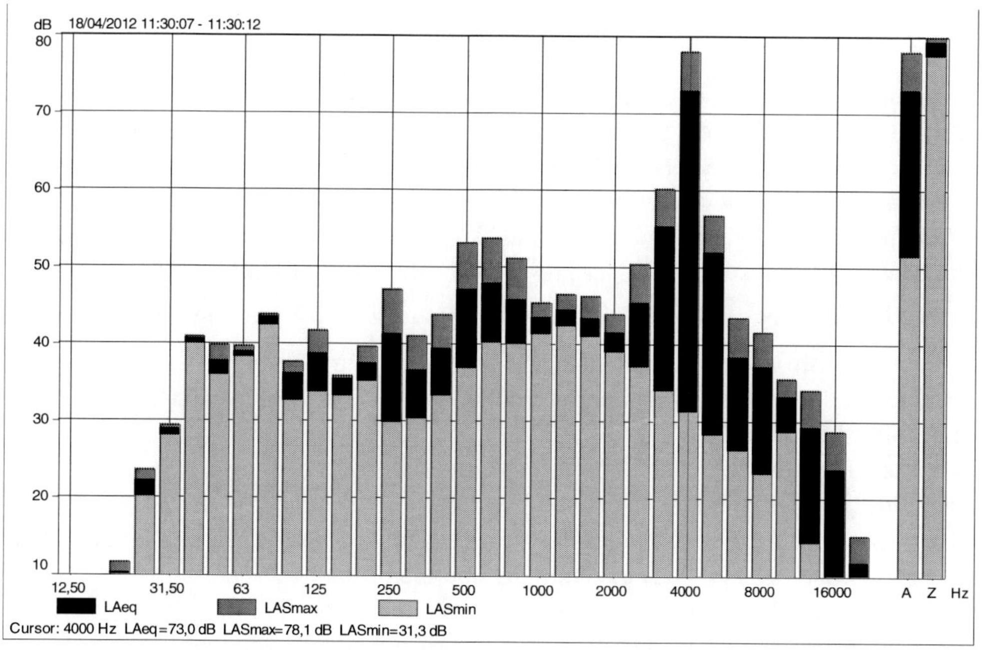

Figure 13. Measured sound spectrum of the ringer. At 4000 Hz, the cursor shows an equivalent sound pressure level of L_{eq} = 73.0 dB(A), maximum sound level of L_{max} = 78.1 dB(A) and minimum sound level of L_{min} = 31.3 dB(A).

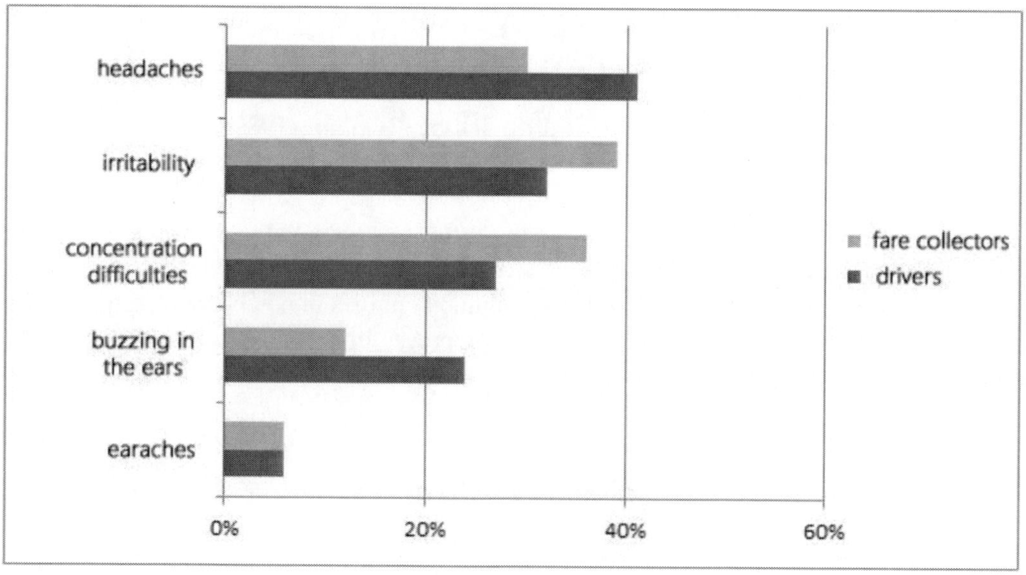

Figure 14. Physical symptoms perceived by bus drivers as a consequence of noise.

Among the interviewees, 21% of the drivers and 21% of the fare collectors reported having difficulties in understanding speech in the presence of noise, while 15% of the drivers and 24% of the fare collectors stated that audiometric testing had indicated hearing loss.

CONCLUSION

The data reported here demonstrate that bus drivers and fare collectors perceive noise as a problem in their workplace environment and are aware of the possible harmful effects it may have on their health. This perception was confirmed by 74% of the drivers and 88% of the fare collectors when they were asked if they believed that noise could be harmful to their health, and particularly to their auditory health.

The statistical analysis of the data presented in this chapter leads to the conclusion that standardized noise exposure levels $L_{EX,8h}$ are directly related to the position of the engine. Hence, buses with front-mounted engines are especially harmful to the health of drivers and fare collectors. Bus drivers (60%) and fare collectors (almost 80%) stated that the engine is the main source of noise.

This study demonstrated that complex environments such as the workplace of bus drivers and fare collectors should not be evaluated based on a single standard. It was also found that subjective assessments (interviews/questionnaires) should be performed jointly with objective evaluations (measurements), since the latter are complementary.

The drivers and fare collectors' perception of noise corroborated the measured noise levels, since 80% of the drivers and 91% of the fare collectors stated that it was necessary to reduce the noise level inside the buses. Another important finding concerning the discomfort for urban bus workers caused by noise is that 77% of the drivers and 88% of the fare collectors feel annoyed and irritated by this physical agent. Only 23% of the drivers and 12% of the fare collectors claimed they felt no discomfort.

Moreover, drivers and fare collectors are exposed to these noise levels 6 hours a day, 6 days a week, during most of the year, a situation that translated into a potential health risk. This finding is consistent with those of several other studies cited in this chapter, which consider bus driving a stressful and unhealthy activity (Gulian, 1989; Evans and Carrère, 1991; Evans and Johansson, 1998; Greiner et al., 1998; Rydstedt et al., 1998; Kompier et al., 2000; Tse et al., 2006).

A possible administrative solution to reduce the occupational noise exposure of bus drivers and fare collectors would be to reduce their daily working hours. The technical solution to the problem would be to build rear-mounted engines in newly manufactured buses. Another important technical solution would be to increase investments in vehicle preventive maintenance aimed at noise and vibration control.

ACKNOWLEDGMENTS

The authors are indebted to the occupational physician Dr. Deisi Mattos Silva and the drivers and fare collectors that participated in the interviews, without whose help this study would have been impossible. The authors gratefully acknowledge the German Government, through the German Academic Exchange Service – DAAD (Deutscher Akademischer Austauschdienst) and the Brazilian Government, through the National Council for Scientific and Technological Development – CNPq, for their financial support, which enabled the purchase of the sound level meters and software used in this study. Authors wish to thank

Prof. Dr. Carolina Arruda Freire for useful suggestions concerning data treatment and analysis.

REFERENCES

Brown AL. Exposure of the Australian Population to Road Traffic Noise. *Applied Acoustics*, 1994, 43: 169-176.

Brazilian Occupational Health Standard NHO-01, 2001, Brazilian Ministry of Labor and Employment (in Portuguese).

Becker T, Tamagna A. Avaliação do ruído produzido em ônibus urbanos em Porto Alegre [*Evaluation of noise produced in urban buses in Porto Alegre*], Universidade Federal do Rio Grande do Sul [*Federal University of Rio Grande do Sul*], Brazil (1998).

Brüel and Kjær. *Technical Documentation – Sound Calibrator BK 4231* (Naerum, Denmark: Brüel and Kjær), 1998.

Brüel and Kjær. *Technical Documentation - Integrating and logging sound level meter BK 2238 and BK 2260* (Naerum, Denmark: Brüel and Kjær), 1998.

Brüel and Kjær. *Technical Documentation – Software Protector BK 7825* (Naerum, Denmark: Brüel and Kjær), 1999.

Babish W, Ising H, Gallacher JE, Sweetnam PM, Elwood PC. Traffic noise and cardiovascular risk: The Caerphilly and Speedwell studies, third phase 10-year follow up. *Archives of Environmental Health*, 1999, 54: 210-216.

Brazilian Institute of Geography and Statistics. 2000 Report. Brasília, Brazil, 2000. (in Portuguese).

Cannelli GB, Glück K, Santoboni S. Traffic noise pollution in Rome. *Applied Acoustics*, 1983, 7: 103-115.

Chakrabarty D, Santra SC, Mukherjee A, Roy B, Das P. Status of road traffic noise in Calcutta metropolis, India. *J. Acoustic. Soc. Amer.*, 1997, 101: 943-949.

Coensel BD, Muer TD, Yperman I, Botteldooren D. The influence of traffic dynamics on urban soundscapes. *Applied Acoustics*, 2005, 175: 194-66.

Diesendorf M. Urban Transportation in the 21st Century. *Environmental Science & Policy*, 2000, 3: 11-13.

Diniz FB, Zannin PHT. Noise impact by electrical energy substations in the city of Curitiba, Brazil. *Sci Total Environ.*, 2004: 328: 23–31.

da Paz E, Ferreira AMC, Zannin PHT. Estudo comparativo da percepção do ruído urbano [*Comparative study of the perception of urban noise*]. *Revista Sáude Pública*, 2005, 39 (3): 467-472 (in Portuguese).

Ewans GW, Carrère S. Traffic Congestion, Perceived Control, and Psychophysiological Stress among Urban Bus Drivers. *Journal of Applied Psychology*, 1991, 76 (5): 658 – 663.

Ewans GW, Johansoon G. Urban Bus Driving: An International Arena for the Study of Occupational Health Psychology. *Journal of Occupational Health Psychology*, 1998, 3(2): 99 – 108.

Fonseca JC. Avaliação da exposição ao ruído dos motoristas e cobradores de ônibus urbanos de São Paulo [*Evaluation of noise exposure of urban bus drivers and fare collectors in*

São Paulo], Fundacentro – Fundação Jorge Duprat Figueiredo de Segurança e Medicina do Trabalho [*Jorge Duprat Figueiredo Foundation for Occupational Hygiene and Safety*], São Paulo, 1993.

Fundacentro. NHO-01: Norma de Higiene Ocupacional, avaliação da exposição ocupacional ao ruído [*Occupational Hygiene Standard, assessment of occupational noise exposure*]. São Paulo, 1999.

Griffiths ID, Langdon FJ. Subjective response to road traffic noise. *Journal of Sound and Vibration,* 1986, 16: 32-8.

Guidotti T, Cottle M. Occupational health problems among transit workers. *Public Health Review*, 1987, 15: 29 – 44.

Gulian E, Mathews G, Glendon AI. Dimensions of driver stress. *Ergonomics*, 1989, 32: 585-602.

Greiner BA, Krause N, Fisher JM. Objective Stress Factors, Accidents, and Absenteeism in Transit Operators: A Theoretical Framework and Empirical Evidence. *Journal of Occupational Health Psychology*, 1998, 3(2): 130 – 146.

Goelzer B, Hansen CH, Sehrndt G. Occupational Exposure to Noise: Evaluation, Prevention and Control (*Dortmund/Berlin: Publication Series from the Federal Institute for Occupational Safety and Health. Document published on behalf of the World Health Organization*), 2001.

Guedes ICM, Bertoli SR, Zannin PHT. Influence of urban shapes on environmental noise: A case study in Aracajú, Brazil. Science of the Total Environment, 2011, 412-413: 66-76.

Herbst K. Brazil's Model City. *Planning*, 1992, 58(9): 24 – 27.

Kumar K., Jain KV. A Study of Noise in Various Modes of Transport in Delhi. *Applied Acoustics,* 1994, 43: 57-65.

Kompier MAJ, Van der Berg AM, Aust B, Siegrist J. Stress Prevention in Bus Drivers: Evaluation of 13 Natural Experiments. *Journal of Occupational Health Psychology*, 2000, 5(1): 11 – 31.

Korfali SI, Massoud M. Assessment of community noise problem in the greater Beirut area, Lebanon. *Environmental Monitoring and Assessment*, 2003, 84: 203-218.

Khilman T. Noise pollution in cities, Curitiba and Göteborg as examples. *Proceedings of the Seminar – Environmental Aspects of Urbanization – Seminar in Honor of Dr. Mostafa Kamal Tolba*, Gothenburg, Sweden, on CD, 2004.

Latance Júnior S. Desmistificando o tabu do ruído dos motoristas de ônibus de São Paulo [*Demystifying the noise taboo of bus drivers in São Paulo*]. *Revista CIPA* [*CIPA Journal*], (1999) (in Portuguese)

Morillas JMB, Escobar VG, Sierra JAM, Gómez RV, Carmona J T. An environmental noise study in the city of Cáceres, Spain. *Applied Acoustics*, 2002, 63: 1061-1070.

Norma Regulamentadora NR-15 – Atividades e operações insalubres, Anexo 1 [*Regulatory Standard NR-15 – Unhealthy activities and operations, Appendix 1*]. Segurança e Medicina do Trabalho [*Safety and Occupational Medicine*], Editora Atlas, São Paulo; 68th ed., 2011 (in Portuguese)

Norma Regulamentadora NR-17 – Ergonomia, Segurança e Medicina do Trabalho [*Regulatory Standard NR-17 – Ergonomics, Safety and Occupational Medicine*], Editora Atlas, São Paulo; 68th ed., 2011 (in Portuguese)

Portela BS, Zannin, PHT. Analysis of factors that influence noise levels inside urban buses. *Journal of Scientific & Industrial Research*, 2010, 69: 684-687.

Rydstedt LW, Johansson G, Ewans GW. The Human Side of the Road: Improving the Working Conditions of Urban Bus Drivers. *Journal of Occupational Health Psychology*, 1998, 3(2): 161 – 171.

Smith N, Hensher D. The future of exclusive busways: the Brazilian experience. *Transport Reviews*, 1998, 18(2): 131 – 152.

Skinner CJ, Grimwood CJ. The UK noise climate 1990-2001: population exposure and attitudes to environmental noise. *Applied Acoustics*, 2005; 66: 231-245.

Tse JLM, Flin R, Mearns K. Bus driver well-being review: 50 years of research. *Transportation Research Part F*, 2006, 9: 89–114.

URBS – Metodologia de cálculo da tarifa técnica *[Methodology for calculating the technical fare]*. Curitiba Municipal Government (accessed on 06 Jan 2012 - www.urbs.curitiba.pr.gov.br) (in Portuguese).

Zannin PHT, Calixto A, Diniz FB, Ferreira JA, Schuhli RB. Incômodo causado pelo ruído urbano à população de Curitiba, PR [*Annoyance caused by urban noise to the population of Curitiba, PR*]. *Revista de Saúde Pública*, 2002, 36: 521-524 (in Portuguese).

Zannin PHT, Diniz FB, Calixto A, Ferreira JAC. A survey of urban noise annoyance in a large Brazilian city: the importance of a subjective analysis in conjunction with objective analysis. *Environmental Impact Assessment Review*, 2003, 23: 245 – 255.

Zannin PHT. Occupational noise in urban buses. *International Journal of Industrial Ergonomics,* 2006, 36 (10), 901-905.

Zannin PHT. Exposure to noise inside transit buses in Curitiba: measurements and passenger attitudes. *32nd Meeting of the German Acoustical Society*, March 20–23, Braunschweig, Germany, 2006.

Zannin PHT. Occupational noise in urban buses. *International Journal of Industrial Ergonomics*, 2008, 38 (2): 232-237. (This paper is a reprint of the Zannin (2006) paper. The paper was reprinted in a special edition of the Int. J. of Ind. Ergonomics in 2008).

Zannin PHT, Sant'Ana DQ. Noise mapping at different stages of a freeway redevelopment project – A case study in Brazil. *Applied Acoustics*, 2011, 72: 479-486.

INDEX

F

G

H

I

M

N

O

P

Q

R

S

T

U

V

W